THE JOHN HARVARD LIBRARY

Howard Mumford Jones
Editor-in-Chief

AN ESSAY ON

CALCAREOUS MANURES

By

EDMUND RUFFIN

Edited by J. Carlyle Sitterson

THE BELKNAP PRESS OF
HARVARD UNIVERSITY PRESS
Cambridge, Massachusetts

1 9 6 1

CONTENTS

AN ESSAY ON CALCAREOUS MANURES

Appendix

Edmund Ruffin,
Agricultural Reformer and Southern Radical

With the publication in 1832 of his first book, *An Essay on Calcareous Manures*, Edmund Ruffin, an obscure Virginian, initiated an era of agricultural reform in the Old South.[1] Within a decade, he had become the most influential agricultural leader in the region and one of the great figures in American agricultural history. By 1850, his work had resulted in the transformation of the economy of the upper South from poverty to agricultural prosperity.

Edmund Ruffin was one of the most remarkable men produced by the ante-bellum South. He had two quite different careers — one as the region's foremost agricultural reformer, the other as a leading Southern radical bent upon taking the South out of the Union. Although it is true that Ruffin's economic views influenced his political radicalism, these two careers were quite separate and distinct. The peculiarities of Ruffin's character would not permit him to attain satisfaction and completion from a constructive career in agricultural leadership. With the same zeal he had devoted to agricultural reform, he turned to a career of political agitation that led eventually to secession and ultimate tragedy. Ironically, history has largely forgotten Ruffin, the agricultural reformer who led Virginia and the upper South along the road of agricultural progress. It remembers Ruffin, the agitator, and the tragedy of a Lost Cause.

Edmund Ruffin was born on January 5, 1794, at the Ruffin family estate, Evergreen, at Coggin's Point on the James River in Prince George County, Virginia. The first of the family, William Ruffin, had come to Virginia in 1666. The family

[1] Although modern readers may find the title amusing and surprising, given the current usage of language at the time of publication the title was not an unusual one. See below p. xiv, note 14.

prospered and by the time of the Revolution was among the leading families of tidewater Virginia. Of Edmund's father, George Ruffin, we know little. He married twice and was the father of seven children. Edmund's mother, Jane Lucas of Surry County, died soon after Edmund's birth, and the rearing of the child fell to his stepmother, Rebecca Cocke. Nor can we with certainty say much of his childhood. He was a frail and sensitive child who read as widely as opportunity permitted, but who apparently took little interest in the day-to-day tasks of plantation operation. With his six half-brothers and sisters, born at close intervals, to demand so much of the love and attention of his stepmother, there was hardly time for her to devote to her sensitive stepchild all the attention he so deeply craved.

After elementary education at the hands of his parents and probably private tutors, at the age of sixteen Edmund entered William and Mary College at Williamsburg. He neglected his studies and utilized his time, for the most part, in idle and wasteful activities. After a brief stay he left college, but not before he had fallen in love with Susan Travis, attractive daughter of a leading Williamsburg family, whom he soon married. Ruffin enlisted as a private in the War of 1812, but he saw no combat service, and after six months of training was discharged. He returned to assume the active direction of the Coggin's Point farm that he had inherited from his father, who had died in 1810. He had now entered upon the career of a Virginia planter to which his heritage of family acres and slaves entitled him.[2]

Although Ruffin was slight of build, a bare five feet, eight inches tall, his face with its gray, deep-set eyes, sharp, well-formed nose, large, straight mouth, and square chin reflected a determination and earnestness that formed the basic core of

[2] *De Bow's Review*, XI (1851), 431–436. See also Avery Craven, *Edmund Ruffin, Southerner* (New York, 1932), pp. 2–4; unfortunately this excellent biography is out of print.

his character. There was little levity in Edmund Ruffin. He took himself and life seriously, and he expected others to do the same. And, as is often the case with such persons, he saw much wrong with things about him and he was determined to set them right. When, as was bound to happen, his efforts sometimes met public indifference or even opposition, he was deeply hurt. The evidences of his generosity and kindnesses to his friends and of his love and affection for his family are abundant. Yet, he could take a slight or criticism to his bosom, nourishing an undying hatred for the offender. In short, Ruffin was a person of sharp feelings; he loved, and he hated; he was rarely indifferent.[3]

Ruffin was a wide reader most of his life, ranging from newspapers and periodicals to the classics, Shakespeare, history, the novels of the nineteenth century, and the latest scientific works. He was also an appreciative critic of the fine arts. Moreover, Ruffin could and did write. A steady stream of correspondence, newspaper articles, diaries, pamphlets, and books flowed from his pen. Two considerable collections of Ruffin's writings remain. One, a sizable group of his letters and a manuscript autobiography entitled "Incidents of My Life," is in the Southern Historical Collection at the University of North Carolina Library, and the other, his diary (1855–1865), is at the Library of Congress.

The young planter had to face necessities, and he spent the early years at Coggin's Point in strenuous efforts to put his plantation on a profitable basis. At this time Ruffin began the agricultural experiments that were to contribute so significantly to the agricultural revolution of the upper South. Although only after later disappointments in other fields did he turn to a career of agricultural reform, these early efforts at practical farming, as well as the inquiring quality of his mind, were responsible for first interesting him in the possibilities of scientific agriculture.

[3] Craven, *Ruffin*, pp. 4–25.

It has been frequently noted that politics was the avocation of the gentleman of the old South. The ante-bellum South reserved its highest esteem and fondest accolades for the political leader, not the writer, artist, or scientist. Once Ruffin had put his plantation and personal affairs in order, he naturally turned to politics, where he could exercise his talents in the service of his state. In 1823, at the age of twenty-nine, he was elected to a four-year term in the Virginia State Senate, representing the district that included the counties of Sussex, Surry, Southampton, Isle of Wight, Prince George, and Greenville. As a member of the State Senate, Ruffin diligently attended Senate sessions and conscientiously performed his services on a number of committees, including the important internal improvements committee.[4] But Ruffin's venture into political life was not a happy one. In the first place, he was not a good speaker and consequently could not attain leadership by that route. Nor, apparently, was he effective in the art of persuasion in committees and in personal contacts with other members. Disappointed in his failure to achieve leadership and resentful of criticism from some of his constituents, he resigned in 1826 before the end of his term.

This was the first major turning point in Ruffin's career and was to open the most constructive and useful phase of his life. In his own mind, however, he had failed in the area of public life in which men of his station usually sought preferment and that was held in greatest esteem by his state and section. He himself wrote later that "few persons would have been more gratified by being so placed [in political stations], and very few young men read more, or felt more interest, on the subjects of government and political economy."[5] But Ruffin buried his disappointment in a tireless effort to make his own plantation more profitable and to chart the way to agricultural progress in his native state.

[4] *Ibid.*, pp. 38–41.
[5] Quoted in *ibid.*, pp. 40–41.

When Ruffin returned to Coggin's Point in 1826 to resume his efforts to make planting pay on the worn-out lands of a tidewater Virginia plantation, he was a far more experienced farmer than the young man of seventeen who had begun this task thirteen years earlier under far different circumstances. Since colonial days a one-crop agriculture based upon tobacco culture had prevailed in eastern Virginia. With capital and labor scarce, and land abundant and cheap, this practice had resulted in the development of large land units and in cultivation of the soil until it had lost its fertility. Then, new land was cleared and the process was repeated. This "mining of the soil" resulted in an impoverishment of the soil that by the early nineteenth century had left much of Maryland and Virginia a region of worn-out land, low property values, an excess supply of Negro slaves, and limited economic opportunities. Virginians by the thousands, seeing only a dismal future at home, emigrated to newer states of the South and Middle West. The rate of population growth dropped from thirty-eight per cent in 1820 to less than fourteen per cent in 1830, and then to a mere two per cent in 1840. As Ruffin himself later described it, "All wished to sell, none to buy." So poor and exhausted were his lands and those of his neighbors that they averaged only ten bushels of corn per acre and even the better lands a mere six bushels of wheat.[6]

True, enlightened planters had preached crop rotation and diversification, and John Taylor, the able planter of Caroline County, Virginia had published in 1813 his notable essays on agricultural practices, *Arator*.[7] Indeed, Ruffin himself not only read and studied Taylor's *Arator* but tried to put into practice Taylor's advice on deep plowing, crop rotation, composting, and the restoration of worn-out lands by inclosing without grazing. Unfortunately, the manure Ruffin applied had little effect on the first year's planting and virtually none

[6] *Farmers' Register*, VII (1839), 659–667.
[7] Seventh edition reprinted in *Farmers' Register*, VIII (1840), 703f.

on the second. His attempt to restore soil fertility by planting clover failed, for his efforts to get good stands on his poor soil were unsuccessful. He persisted for four or five years, but Taylor's methods "proved either profitless, entirely useless or absolutely and in some cases greatly injurious." [8]

Little was then known in the United States of soil chemistry. While the use of putrescent manure [9] might restore the original fertility to a given piece of land, if the land were originally lacking in mineral ingredients essential to fertility, the land would still yield sparingly. So discouraged had Ruffin become that he contemplated selling his estate and emigrating Westward. At this time he came across a copy of Sir Humphrey Davy's *Elements of Agricultural Chemistry* (London, 1813). Ruffin knew nothing of chemistry but he was struck by Davy's statement that "If on wash[ing] a sterile soil it is found to contain the salts of iron, or any acid matter, it may be ameliorated by the application of quick lime." Some soils that were apparently of good texture contained sulphate of iron and could be made fertile by the application of lime, which converted the poisonous substance into manure. [10]

Could the soils of Coggin's Point be sterile because of the existence of sulphate of iron? But tests demonstrated that they did not after all contain sulphate of iron. Ruffin then noted that plants known to contain acid, especially sheep-sorrel (*rumex acetosa*) and pine, grew luxuriantly in soils unsuited

[8] *DeBow's Review*, XI (1851), 432. See also Edmund Ruffin, *An Essay on Calcareous Manures*, p. 27. All page references to Ruffin's *Essay* are to the John Harvard Library edition.

[9] Ruffin used the term putrescent manures to refer to those forms of vegetable and animal matters that in the process of decay and decomposition furnished soluble food to plants. See Ruffin, *Essay*, p. 10n.

[10] Davy, *Agricultural Chemistry* (Philadelphia, 1821), p. 141. Ruffin's dependence upon Davy is clearly evidenced by his many citations of Davy's work. It was not until 1841 that the American edition of Justus Liebig's *Chemistry in Its Application to Agriculture and Physiology* appeared. In this work Liebig pointed out that certain mineral elements that were supplied by the soil were necessary to plant growth and that when certain constituents were lacking in the soil they might be supplied through the application of mineral manures.

to cultivated crops. He surmised that an excess acidity was responsible for soil sterility and that this acidity prevented the soil from receiving the benefits of vegetable manures.[11] Although all his tests failed to indicate the presence of any mineral acid, he noted that the poor soils contained no calcium carbonate.[12] Conversely, he observed that those small areas of his land which were shelly and, of course, calcareous, contained no pine or sorrel. Accordingly, Ruffin determined to experiment with improving his soil by the addition of calcareous earth (i.e., containing calcium carbonate) or marl (high in its lime content). Marl had long been used successfully in European agriculture, and had occasionally been used in Virginia since colonial days but without notable success.

In February 1818 under his direction his slaves dug marl (containing about 33⅓ per cent calcium carbonate) from one of the beds of fossil shells which underlay much of the land of tidewater Virginia and applied it at the rate of 125 to 200 bushels to two or three acres of newly cleared land. In the spring he planted it in corn. As soon as the plants were a few inches high, their superiority over the plants on adjacent land was apparent. When the corn was gathered in the fall the yield was fully 40 per cent above that of his other land. Ruffin's enthusiasm was unbounded. In October he presented the results of his experiment in a short paper to the agricultural society of Prince George County, and over the next years he applied marl extensively to more and more of his acreage with gratifying results. In 1821 he presented his experiments in a paper entitled "On the Composition of Soils, and their Improvement by Calcareous Manures" in the *American Farmer* (III, 313–320), published in Baltimore. John Skinner, the editor, called Ruffin's paper "the first systematic attempt . . .

[11] Vegetable manures include any plant materials used to fertilize soil. When vegetable and animal manures are mixed together and lime is added to assist in the process of decomposition, it forms what is commonly called compost.
[12] The vegetable or humus acids were not discovered until a much later date.

wherein a plain, practical, unpretending farmer . . . has undertaken to examine into the real composition of the soils which he possesses, and has to cultivate." Skinner considered the paper so important that he published an extra edition of that issue of the *Farmer* for free distribution to farmers.[13]

Ruffin had never had any formal training in chemistry, and it would not be correct to call him a scientist. Essentially he was a gentleman farmer, in many respects like thousands of others in the ante-bellum South, but with important differences. Unlike most, he read widely, had an insatiable curiosity, and employed a scientific approach in his efforts to solve the agricultural problems of his region. One cannot but be struck by the laboratory methods used by this "practical farmer" in analyzing the mineral contents of the soil and his realization of the importance of accurate records of the results.

In 1826 Ruffin expanded his paper with the intention of publishing it in book form. However, after his close friend, Thomas Cocke, made unfavorable comments on its style, Ruffin put the manuscript away until 1832, when he finally published it, with very little change, under the title, *An Essay on Calcareous Manures*.[14] The small edition of 750 copies was soon sold. Briefly, Ruffin maintained that: (1) the capacity of soils for being enriched by putrescent manures is only equal to their original fertility prior to cultivation; (2) the soils of the Atlantic slope of Virginia, and possibly of other Southern states, were lacking in calcium carbonate; (3) most of the poor soils in this area contained vegetable acid, which was a cause of the sterility of the soil; and (4) the application of calcium carbonate would neutralize the acid, thereby making it possible to improve the land and cultivate it profitably.[15]

[13] *American Farmer*, III (1821), 313.

[14] At the time of Ruffin's writing, the term manure had not come to be used generally as synonymous with animal manure, but included any material used to fertilize land. Thus the term "calcareous manures" accurately described a substance containing calcium carbonate, such as marl, which could be used for soil improvement.

[15] Ruffin, *Essay*, pp. 21–22.

Ruffin did not contend that the use of marl alone was a panacea for the agricultural problems of his region. He was fully cognizant of and a practitioner of the agricultural methods recommended by Taylor and other agricultural reformers. Rather, he emphasized that marl would prepare the soil so that the application of organic manures [16] would be effective to soils on which they had previously been of little value. In other words, for certain soils the application of marl was the indispensable first step. Marl, he contended, corrected the natural acidity of the soil and assisted in the preservation of the gaseous products that accompanied the decomposition of organic manures.

Little was then known of the action of soil bacteria, and it is remarkable that Ruffin's experiments and deductions foreshadowed later discoveries that certain nitrifying organisms act only in neutral or alkaline soils. Indeed, it was Ruffin's concept of fertility as a dynamic and not a static condition of the soil that made him a precursor of the modern science of soil chemistry, and may well justify calling him the father of soil chemistry in America. Ruffin accepted the view that soil fertility was dependent upon the chemical contents of the soil, but he realized what was to be known conclusively only years later, that the chemical composition of soil, and consequently its fertility, was subject to change as a result of organic action in the soil. Ruffin's *Essay* is a landmark in the history of soil chemistry in the United States, though it has been largely superseded by the advances of knowledge in this field.

Unlike Taylor's *Arator*, with which it is sometimes compared, Ruffin's work did not treat of agricultural problems in general, much less of the relationship of agriculture and political policies to which much of *Arator* is devoted. Rather Ruffin attempted to treat, as thoroughly as existing scientific

---

[16] Organic manures include any plant matter, such as leaves and other plant growth, or animal matter, such as farmyard and stable manure, used to fertilize soil.

knowledge permitted, a single agricultural problem. His wide acquaintance with the contemporary literature in agricultural chemistry is clearly indicated by his many citations of works by English, French, and Scotch writers on agriculture.

After its first publication in book form in 1832, Ruffin's *Essay* appeared in four subsequent editions, the fifth appearing in 1852. During the twenty-year period it grew from a seven-page article in the *American Farmer* in 1821 to a book of 242 pages in 1832, and finally to 493 pages (with appendices) in 1852. As the *Essay* was enlarged, Ruffin added directions for finding marl beds, analyzing the composition of marl, data on the cost of marling land, and the progress of marling in Virginia since 1818.[17]

Ruffin's *Essay* was widely and favorably reviewed in agricultural journals in the United States and Great Britain. Although some reviewers took issue with Ruffin on minor points, without exception they regarded his major thesis as sound. One reviewer called it a "masterpiece," another "an original work of great merit" that should be read by every farmer, and an English reviewer declared that Ruffin had performed a "very imporant service to the scientific agriculturist in this country, as well as America." [18] As late as the end of the nineteenth century it was called by an expert in the United States Department of Agriculture "probably the most thorough piece of work on a special agricultural subject ever published in English." [19]

There can be no doubting the fact that Ruffin's *Essay* was

[17] The second and third editions were printed as supplements to the *Farmers' Register* in April 1835 and Dec. 1842. The fourth edition was printed in 1844 for special distribution in South Carolina. It is difficult to say with certainty how many copies of the *Essay* were printed and distributed, but it is likely that the five editions totalled about five or six thousand copies.

[18] Reviews reprinted in *Farmers' Register*, II (1834), 632; III (1835), 717–721; IV (1836), 95–104.

[19] W. P. Cutter, "A Pioneer in Agricultural Science," in U.S. Dept. Agri., *Yearbook, 1895* (Washington, 1896), p. 500.

widely read in the ante-bellum South and that it had a profound influence on agricultural practices. In the 1840's and
1850's, marl was extensively used by progressive farmers
where beds were readily available in Delaware, Maryland, Virginia, and to a lesser extent in North and South Carolina. The
results were impressive. Ruffin almost doubled his wheat yields
at his Coggin's Point farm — from less than six bushels per
acre to more than ten. Later at his Marlbourne plantation,
with marling and drainage Ruffin increased his wheat yield
from less than fifteen bushels per acre in 1844 to twenty in
1848. Scores of other farmers experienced similar success. Unfortunately the supplies of available marl were limited and the
high cost of transporting it prevented profitable use over wide
geographical areas. Indeed, so evident were the improvements
attributable to the *Essay* and Ruffin's other activities in support of agricultural reform that Governor Joseph Johnson of
Virginia in his annual message to the legislature in January
1852 called attention to the substantial increase in the value of
lands in the tidewater district in these words:

This remarkable and gratifying change in the value of these
lands cannot be attributed, to any great extent, to benefits result-
ing from the works of internal improvement; for thus far these
improvements have been chiefly confined to other sections of the
State. And in vain will we look for a solution of this problem,
unless we remember that for several years past, the enterprising
citizens of this section of the state have been devoting themselves,
. . . to the subject of agricultural improvement; and by a proper
application of compost, marl and other manures, and the use of
other means which a knowledge of this branch of education has
placed at their command, they have redeemed and made productive and valuable, lands heretofore worn out by an improper
mode of cultivation, and consequently abandoned by the farmer
as worthless and unfit for agricultural purposes.[20]

The significance of the problem of soil exhaustion in the
ante-bellum South, though familiar to specialists, is not always

[20] *Journal of Senate of Virginia* (Jan. 1852), p. 14.

appreciated by the general reader. Since the aboriginal Indians were not characteristically an agrarian people, when the European settlers first began to farm the new land, they found much of the soil of the New World amazingly rich. Inasmuch as they knew little about soil renewal, the first farmers — and the tobacco growers of colonial Virginia were characteristic but not unique in this respect — assumed that this fertility would never cease. When experience proved that it did, and a given piece of land was exhausted, they cleared new land, abandoned the old, and repeated the process. By the time Ruffin turned his attention to soil chemistry, the economy of eastern Virginia had suffered severely as a result. A like process was driving farmers North and South steadily westward in the search for virgin land they could again despoil. Hence it is that although Ruffin's immediate impact was upon his neighbors and upon the South, his book takes its place among the national classics of American agriculture.

Agricultural reform did not come easy in the ante-bellum South. Statesmen and agricultural leaders, it is true, were quick to utter their praises. Practical farmers, then as now, however, were slow to change their ways of doing things. Ruffin saw that some means must be found to get knowledge of the latest and best agricultural practices to the farmers of his state and region. The success of the first edition of his *Essay* encouraged him to undertake the publication of an agricultural journal. In June 1833, the first number of the *Farmers' Register*, a monthly periodical of sixty-four pages, was issued from his Coggin's Point plantation, which he had designated Shellbanks. "With this publication," Ruffin wrote, "was begun a new & distinct era of my life." The time was opportune for such a venture, and the response was gratifying. After the first year, when he moved his place of publication to Petersburg, Virginia, 1,118 subscribers in eighty-five Virginia counties and towns and 115 from fifteen other states and the District of Columbia had indicated their willingness to pay the subscription rate of five dollars a year. Within three years, subscriptions had increased to about 1,400.[21]

Ruffin announced the policies of the *Register* in an advance prospectus. There were to be no advertisements. Articles from practical farmers would be especially welcomed. There was to be no discussion of politics. The entire contents of the magazine were to be devoted to the improvement of agriculture.

Ruffin dominated his journal as did no other ante-bellum agricultural writer and himself wrote nearly half the material in its pages. His powerful editorials covered almost all phases of agriculture. Marl and its uses, understandably a favorite subject for Ruffin, were discussed repeatedly. In addition, he reprinted selections from the works of the leading agricultural writers of Europe and America, including in full John Taylor's *Arator*, the *Westover Manuscripts*, Davy's *Elements of Agricultural Chemistry*, and the second and third editions of his own *Essay on Calcareous Manures*.

The quality of the *Farmers' Register* was manifest from its first issues, and for ten years the *Register* maintained this high standard, and was unrivalled in quality by any other agricultural periodical in the country. The numerous articles by farmers contained in its pages are a good indication of the agricultural changes of the 1830's and early 1840's. It is not too much to say that Ruffin's *Farmers' Register* was the most important single force in producing the agricultural revolution that transformed the agriculture of the upper South from its moribund state in the 1820's to its prosperous condition in the 1850's.

When Ruffin and the *Register* began their work, the agriculture of the upper South seemed doomed. Early agricultural reformers had failed, agricultural societies were drying up, farmers after repeated failures had concluded that emigration Westward was their only hope. The use of marl and the application of other improved methods of agriculture persistently

[21] Ruffin, "Incidents of My Life" (MS), Univ. of N. C. Lib.

advocated in the pages of the *Farmers' Register* had their effect. A profound change occurred in the agriculture of Virginia. Crops increased, land improved in value, and emigration declined markedly. From 1838 to 1850, the land values of tidewater Virginia increased by more than seventeen million dollars. Ruffin's neighbors in Prince George County in 1843 expressed their gratitude to one who had "devoted his time, his talents, his money and industry, in endeavoring to convince us by practice, as well as by luminous productions on calcareous manure, how we might use marl, [and] reclaim the barren fields with which our county abounds." [22]

It would be interesting to speculate on the future career of Ruffin and the further influence of the *Farmers' Register* on Southern agriculture had the *Register* continued its successful publication throughout the 1840's and 1850's. However, as the *Register* became most influential, the seeds were sown that were to eventuate in its demise. Even had Ruffin been without pronounced political views that grew stronger with the years, it would have been difficult to keep political issues out of the pages of the *Register*. Farmers and politics have been bedfellows throughout American history. As important political and economic issues arose that Ruffin considered to be significant for his state and region, how could he view them as unrelated to agriculture, the principal economic interest of the South? His identification with the Southern attack on the protective tariff created no real difficulty. After all, except for the Louisiana sugar planters, Southern farmers were agreed that the tariff was injurious to Southern agriculture and to the economic interests of the region.

Unfortunately for Ruffin and the *Register*, there was far less agreement on the question of bank reform to which Ruffin directed his attention in the late 1830's. Excessive issues of bank notes and easy credit were important causes of the Panic

[22] Resolutions of Farmers of Prince George County, Oct. 10, 1843, Peyton Bolling, Chairman; copy in Ruffin Papers, Univ. of N. C. Lib.

of 1837 that hit the South as well as the remainder of the
nation. Ruffin himself lost considerable sums by unsound in-
vestments and too ready loans to his friends. He was quick to
conclude that the unsound banking practices and particularly
the issuance of bank notes were injurious to Southern agriculture and that banking reform and agricultural reform were
vitally related. At first, the articles on banking that appeared
in the *Register* discussed with moderation the entire subject
of banking, the issuance of bank notes, and the extension of
credit. But moderation was not a part of Ruffin's nature, and
soon his tone changed markedly. Ruffin lashed out repeatedly
at the "frauds and abuses of the banking interests" and determined "to awaken the members of the agricultural interests
of Virginia, and of the whole Confederacy, to a sense of the
enormous evils . . . suffered, and the system of pillage . . .
still pursued by the banking system of this country." [23]

The reaction was quick to come. Newspaper editors, formerly friendly, criticized the new emphasis of the *Register*.
Readers cancelled subscriptions. But opposition only confirmed Ruffin in his new crusade. On September 4, 1841, he
began publication of the monthly *Bank Reformer*, for which
he chose the motto, *"Nil Utile Quod Non Honestum."* Ruffin's
purpose, as he expressed it, was to furnish information on the
correct principles of banking and to expose the "wrong-doing
and fraudulent practices" of the existing banking system. To
reach the largest possible number of readers, Ruffin not only
sold issues in quantity at $5.00 per two hundred copies, but
distributed them to postmasters with instructions to deliver
a copy to anyone paying the postage.

Nor did Ruffin stop with the *Bank Reformer* and with the
columns of the *Register*. His letters, over the signatures of the
names of the enemies of tyranny in ancient Rome, appeared
in newspapers friendly to his cause. He even went so far as

[23] See *Farmers' Register*, IX (1841), 163–166, 618–619, etc.

to print on the back of bank notes that reached him such statements as the following:

The promise on the face of this note is *Fake*; and the issue of such notes is both a banking and a government *Fraud*, committed on the right and interest of labor and honestly acquired capital.[24]

Ruffin had now attacked some of the most powerful persons and interests in Virginia and the nation. His critics, as well as subscribers to the *Farmers' Register*, were quick to point out that he had put the *Register* into politics. Subscription cancellations mounted, and monetary losses inevitably followed. In 1842 Ruffin found it necessary to cease publishing both the *Register* and the *Bank Reformer*. On April 30, 1842, he wrote: "But with the close of this volume will end the editor's labors for ten years of the best years of his life; and he will no longer obtrude on the agricultural public, services which seem to be so little appreciated, and which have been so little aided by the sympathy of the great body of the members of the interests designed to be served." [25] To cap his disappointment, his carefully worked-out plan for extensive agricultural experimentation under the direction of the State Board of Agriculture was rejected. He resigned from the Board. "Disappointed & mortified — soured with the injustice & ingratitude" of his countrymen, Ruffin, as he had earlier in 1826, turned again to his own private affairs "with a determination never to renew a connection with the public in any form." [26]

By his own agricultural experiments and by the publication of his *Essay* and the *Farmers' Register*, Ruffin had pointed the way to agricultural revival for the South. Nevertheless, another period of his life had ended, an important one for the future of his beloved state and region. His wide renown, his determined views, his zeal for their propagation, and his skill as a propagandist would soon open a new career for him, a

[24] Ruffin, "Incidents of My Life."
[25] *Farmers' Register*, X (1842), 155.
[26] Ruffin, "Incidents of My Life."

career that would bring him great satisfactions at times but that would lead ultimately to catastrophe for the South and a tragic end for Ruffin.

Meanwhile, a brief interlude in Ruffin's career came from an unexpected but welcome source. South Carolina found it increasingly difficult to compete in the raising of cotton with the new Cotton Kingdom in the lower South — Alabama, Mississippi, and Louisiana. Some of the state's leaders despaired of rebuilding the economic fortunes of the state on the basis of cotton planting and resolved to chart a new economic course for South Carolina. Among them was James Henry Hammond, who held that the future of South Carolina, and indeed of the entire upper South, lay in agricultural diversification, commerce, and manufacturing. In 1842, Hammond was elected governor and he had his opportunity to set his state on the path of economic progress. He had known for some time of the work of Ruffin in the agricultural revival of Virginia, and if possible he was determined to bring him to South Carolina to provide the leadership for agricultural reform.

Ruffin happily accepted Governor Hammond's invitation to become agricultural surveyor of South Carolina and early in 1843 he went south to take up his new work. Ruffin's reputation as an agricultural reformer had preceded him, and he was welcomed with enthusiasm and appreciation by planters, political leaders, and publicists. Five hundred copies of his *Essay on Calcareous Manures* were distributed over the state to explain the beneficial uses of marl.

Ruffin immediately set to work to locate marl beds in the state, to visit plantations, and to carry the message of agricultural reform. But he had set himself too arduous a task for so frail a constitution. Following the river courses, tramping the swamps in search of marl beds, and exposing himself to weather were more than his health could stand, and he was forced to resign before a year had passed. But warm and permanent attachments had been formed. Ruffin found Charleston and

South Carolinians to his liking. Regretfully he returned to his native state, determined to stay out of public affairs.

During the years of the *Farmers' Register*, Ruffin had in 1835 moved his family and his printing establishment from Shellbanks on his Coggin's Point farm to Petersburg. He paid little attention to his farming operations, leaving them increasingly in the hands first of overseers and then of his oldest son, Edmund, Jr., who purchased half of the farm in 1839 and the remaining half in 1848. Before the end of his South Carolina sojourn, Ruffin had begun to look for a new plantation. In the fall of 1843 he made plans to move to his new home, Marlbourne, in the valley of the Pamunkey River. Taking note of his imminent departure, his neighbors in Prince George County on December 28, 1843, tendered Ruffin a dinner in appreciation of what he had done to transform their farms from poverty to prosperity. But for Ruffin the thanks came too late. He accepted their tribute and moved to his new home in Hanover County, not, however, without noting that the Virginia papers had failed to print his remarks at the Prince George meeting. Such neglect only strengthened his determination to withdraw from public activities.[27]

His new plantation, Marlbourne, with its almost thousand acres was to be his home for the remaining years of his life. The dwelling house, a two-story rectangular structure, was large and comfortable, and nearby were a kitchen, dairy, meat house, ice houses, laundry, stables, carriage house, overseer's house, and slave quarters. All in all, it was an estate that befitted one of the South's leading planters and its best known agricultural reformer. With a slave force of some forty Negroes, Ruffin set to work to make his new place profitable. The task that faced him was no easy one, for the land was poor and badly in need of attention. For the next five years, Ruffin stayed out of public life. So determined was he to remain out of public life that he declined to accept the office of

[27] *Ibid.*

president of the Virginia State Agricultural Society when an effort was made to organize such a society early in 1845, a refusal that he later termed "ungracious."

With unbounded energy and zeal, he immediately began the "marling" of his poor soil. He obtained marl from the beds on the farm of his neighbor, Carter Braxton, and all the available time of his slave force for the next two years was devoted to spreading marl over some 800 acres. He tried clover, cowpeas, and barnyard manure to enrich his soil, always keeping the most careful record of crop yields. In a region of heavy rainfall and with much swampy land, he soon found it necessary to provide a drainage system. His extensive and well-planned system of covered drains contributed substantially to the success of his crops. Wheat, corn, oats, potatoes, fruits, and melons were all produced in his fields, and a carefully devised crop rotation scheme was developed to assure maximum production. Results were not long in coming. Within a few years his crops were bringing in a sizable financial return, and by the early 1850's he had rebuilt his fortune to what it had been prior to his losses in the years after the Panic of 1837.[28] Thus for a second time, first at Coggin's Point in the twenties and early thirties and now at Marlbourne, Ruffin had demonstrated that good agricultural theory and careful farming practices could together provide a sound financial basis for Southern agriculture.

This was a fruitful period of his life and, except for the death of his wife in 1846, could have been a happy and satisfying one. But Ruffin was not content to live a quiet life on his plantation, apart from the mainstream of events. In 1849, in response to a request, he wrote for the July issue of the *American Farmer* an article on "Farming Profits in Eastern Virginia: The Value of Marl." The article attracted wide attention and Ruffin was soon sending frequent articles to the press on a variety of agricultural topics. In 1851 he accepted

[28] *Ibid.*

an invitation from the editor of the widely circulated *DeBow's Review* of New Orleans to supply biographical data and a photograph for a sketch of him to be included in a "Gallery of Industry and Enterprise." [29] Ruffin was highly pleased, and the complimentary notices the article received in the Virginia press seemed to indicate to Ruffin that at last his state was giving him his proper due. In February 1852 a state agricultural society was finally formed, and Ruffin this time accepted the presidency and zealously worked to put the new organization on an active basis. Over the next years he accepted many of the invitations to speak that came to him from agricultural groups in all parts of Virginia and neighboring states. In 1852, the fifth and final edition of his *Essay on Calcareous Manures* appeared. In 1855, he brought together fourteen of his previously written essays on agricultural topics in a volume entitled *Essays and Notes on Agriculture*.

Ruffin continued his agricultural experiments and wrote articles for the press on a variety of agricultural subjects until 1861, but the course of events in the nation and especially the South in the 1840's and 1850's increasingly channeled Ruffin's interests in other directions. Although Ruffin became an active Southern nationalist and secessionist long before most of his contemporaries, the course of his transformations did not differ substantially from that of other Southern radicals. As a young idealist, Ruffin had considered slavery an evil, but with the growth of abolitionism in the 1830's, he became a defender of the institution. Even so, the ten volumes of the *Farmers' Register* from 1833 to 1843 contain little on the subject of slavery. Ruffin's conversion to an active champion of Southern nationalism in the 1840's was rapid. Although he was a wide reader, he knew little at first hand of any area outside the South. He had been in the North only once. Moreover, Ruffin's life and interests were so predominantly agrarian that it was easy for him to see in Southern agricultural society a

[29] *De Bow's Review*, XI (1851), 431–436.

civilization far superior to the incipient industrial society of the North.

In the early 1840's, Ruffin was too resentful of the slights of his own neighbors to direct his bitterness solely at the North. However, when the Northern anti-slavery extremists proposed to limit the further expansion of slavery in the territories Ruffin concluded that they were bent upon destroying the South and that the South's only recourse was withdrawal from the Union. The Compromise of 1850 between the North and the South, which settled this sectional controversy for the time, was unsatisfactory to Ruffin and other Southern radicals. Ruffin was disappointed that South Carolina did not lead the way for the South by seceding from the Union in 1850, but such a stand was far in advance of the greater part of Southern opinion at that time.

As he received more and more recognition at home, Ruffin's hatred of "Yankees" became more implacable. It is not surprising that one with so ready a pen and such a penchant for the public press as Ruffin should soon enter the fray in defense of the South and Southern institutions. Throughout the 1850's he wrote articles on slavery for various Southern newspapers and periodicals. These were in turn enlarged and printed in pamphlet form under the following titles: *The Influence of Slavery, or its Absence, on Manners, Morals, and Intellect* (1852); *Consequences of Abolition Agitation* (1857); *The Political Economy of Slavery* (1857); *African Colonization Unveiled* (1859); *Slavery and Free Labor Described and Compared* (1859).[30] These pamphlets were printed by the thousands and circulated widely throughout the South. Although containing little new, they were an effective statement of the pro-slavery argument. Ruffin was convinced of the inferiority of the Negro and he contended that the condition of

[30] A useful bibliography of Ruffin's writings is Earl G. Swem, *An Analysis of Ruffin's Farmers' Register, with a Bibliography of Edmund Ruffin*, Bulletin of the Virginia State Library, XI (Richmond, 1918).

the Negro slave in the South was better than that of the free laborer in the North. Moreover, he argued that a slave system provided the basis for a higher civilization and cited the glories of ancient Greece and Rome and the culture of the ante-bellum South as proof.

In 1855, Ruffin gave up the active management of his planta-tion, and while he continued to write articles on agriculture from time to time, he increasingly turned his attention to pub-lic affairs. As the bitterness between the North and South in-creased in the closing years of the decade, Ruffin, with bound-less energy that belied his more than sixty years, determined to join the other Southern radicals in showing the South that its rights could be preserved only by seceding from the Union. He visited Washington to urge Southern leaders to stand firm for Southern rights, and he traveled to North Carolina, South Carolina, and Alabama to exchange views with Southern ex-tremists and to convert the timid. In May 1858, Ruffin and William L. Yancey, the eloquent Alabama radical, launched the "League of United Southerners" to awaken the mind of the South and prepare it for secession, but the South was not yet ready for radical action and the activities of the radicals drew sharp criticism.

As Ruffin despaired of the South's facing up to what he considered to be the realities of the time, events played into the hands of extremists in the North and South. On the night of October 16, 1859, John Brown and a small band of men seized the Federal arsenal at Harper's Ferry, Virginia, as the first step in a plan to free the Negro slaves in the South. When the news reached Ruffin on the morning of October 19, his reaction was immediate. He was convinced that his predictions were at long last coming true and that Northern abolitionists had planned a slave insurrection to destroy Southern society.

After presenting a petition to a Richmond meeting of the Southern Rights Association urging every Southern state to prepare to defend the people against Northern assaults, he went to the little village of Harper's Ferry to be present at the execution of John Brown. The execution was to be open only to the military guard, but Ruffin secured permission to serve as one of the color guard in a company of cadets of the Virginia Military Institute. On the morning of December 2, in a long grey overcoat with plain parade cap and carrying a musket, this stern old Virginian with his long white hair falling to his shoulders watched from the ranks as Brown ascended the scaffold. When the deed was done, Ruffin wrote, not without a note of admiration, that Brown appeared completely fearless and insensible to danger and death.[31]

From the John Brown raid until the outbreak of the Civil War, Ruffin worked ceaselessly to bring about the secession of the Southern states. In his view, only by a dissolution of the Union and the formation of a new Confederacy could the South be saved. In the spring of 1860 he wrote *Anticipations of the Future*, which first appeared in part as letters to the Charleston *Mercury*, and was published in book form in the fall of 1860. His *Anticipations* pictured the course of events in the United States from 1864 to 1869, and prophesied the triumph of abolitionism, the final secession of the South, a short but ruinous war, a break between the West and North, and a resurgent South on the road to a new and greater prosperity.

As soon as the news of Lincoln's election had been received, Ruffin hastened southward to Columbia, where the South Carolina legislature was about to call a convention to consider "the value of the Union." In the days following the convention call, Ruffin enthusiastically accepted the plaudits of South Carolina radicals. He then moved on to Georgia to exert what influence he could upon the legislature there to call a convention. Disappointed to find little prospect for immediate action in Georgia, he returned to his native Virginia only to find the conservatives and the advocates of compromise in control. A few weeks later he returned to South Carolina, where as a

[31] Edmund Ruffin to his sons, Dec. 3, 1959, Ruffin Papers.

guest in the convention he witnessed the adoption of the ordinance of secession. From South Carolina, he turned southward to Florida to urge immediate secession.

When he returned to his native Virginia, he found the conservatives still in control. This was too much for Ruffin. Disappointed in the failure of his native state to join the states of the lower South in secession, he turned southward to Charleston, his adopted home. Certain that the Lincoln government would not compromise and that war was inevitable, he wanted to be in Charleston when the fighting began.

Early in April 1861, the Federal government made plans to send supplies to Major Anderson at Fort Sumter. Ruffin, now sure that the struggle would begin any day, joined the other recruits who crowded the boat to reinforce the troops on Morris Island in the harbor. The new recruit was welcomed enthusiastically by the men and on April 10 was unanimously elected a member of the Palmetto Guards. When the order was given to open fire on Sumter on the morning of April 12, Ruffin was notified by his company commander that he had been chosen to fire the first shot. When the signal came at 4:30 on the morning of the twelfth, Ruffin was at his post ready to fire. The shell hit the fort, and the other guns quickly opened fire. Ruffin stayed at his post all that day and the next until the news came of the unconditional surrender.[32]

Ruffin's career had reached a dramatic climax. He had labored long to bring about the dissolution of the Union and the formation of a Southern Confederacy. Now in the evening of his life it had been his privilege and reward to strike the first blow for Southern independence. Ruffin was a hero in Charleston, and his popularity in the South reached a new height.

The firing on Sumter and Lincoln's call for troops to crush

[32] There is some doubt as to who fired the first shot since several batteries fired almost simultaneously. In addition to Ruffin's claim, a good case can be made for both Captain George S. James and Lieutenant W. Hampton Gibbes.

the "rebellion" turned the tide to secession in the upper South. When he received the news of the secession of Virginia, Ruffin ended his voluntary exile and returned home. But too much was happening for Ruffin to remain quietly at home directing plantation affairs. In July he joined the South Carolina Palmetto Guards as they moved through Virginia toward Fairfax Court House. When the orders came to fall back toward Manassas Junction, Ruffin tried with all his might to stay in the ranks. Such a pace was too much for the old man of sixty-seven, and he accepted a ride on one of the artillery caissons. When the fighting began a few days later, Ruffin had the honor of firing a cannon shot at one of the retreating Yankee lines. To his disappointment, the shot felled only three of the enemy.

Ruffin was sure that the war would end soon and he went to Beechwood, Edmund Ruffin Jr.'s home on the Coggin's Point farm. His health was now broken. The exposure and the wear and tear of the campaign had had their effect on his frail constitution. Now that he could no longer contribute to the war against the Yankees, he longed for death. As the months lengthened into years, the early victories were followed by defeats and devastation. Beechwood was plundered, stripped, and ravaged by Yankee troops in the summer of 1862. Ruffin and the family moved to Marlbourne and later to a farm in Amelia County, southwest of Petersburg, where the danger was not so great. With the passing of each month, Edmund Ruffin's world more and more fell apart. One of his grandsons was killed in battle. His daughter Mildred died in Kentucky. His son Julian was killed in battle in May 1864 at Drewry's Bluff. As the year 1865 opened, Ruffin knew that the end was in sight.[33]

When Lee surrendered on April 9, Ruffin realized that more than the Confederate States of America had come to an end.

[33] The closing years of Ruffin's life are described in detail in his "Diary, 1861–1865" (MS), Lib. Cong.; see also Craven, *Ruffin*, pp. 205–259.

His world had disintegrated. The South that he had loved was no more. In the weeks that followed, Ruffin contemplated both the past and the future. His thoughts went back to that February day in 1840 when his closest friend, Thomas Cocke, old and weakened by illness, had taken his own life. After the death of Ruffin's own father, Thomas Cocke had been almost like a father to him. At the time, Cocke's suicide had made a profound impression on Ruffin, and he had written a long and detailed account of the occurrence, "Closing scenes in the Life of Thomas Cocke," and filed it in his papers. Perhaps he reread it on one of those days in May 1865. Were not some of the arguments Cocke had used to justify his own death equally applicable to him? His life was of no more use to him or to his family. He would only be a burden to the ones he loved.

Basically the character structures of Cocke and Ruffin were similar. Both were subject to chronic depression. Ruffin thought that Cocke, who lived much of his life as a recluse, despised his fellow man. For his part, Ruffin's insecurity began in his early life, when after the death of his mother he probably failed to receive from his stepmother the amount of love and attention his sensitive nature needed. Throughout his life, Ruffin complained to the point of despair of the failure of his contemporaries to appreciate his services. Though quick to criticize others, he was deeply wounded when criticism came his way. Fundamentally, despite his remarkable achievements and contributions, Ruffin's adjustment to reality was incomplete. When the world he loved and wanted disappeared, he was unable or unwilling to accept a substitute. On the last page of his diary, he paid his last respects to the foe:

I here declare my unmitigated hatred to Yankee rule — to all political, social and business connections with the Yankees and to the Yankee race. Would that I could impress these sentiments, in their full force, on every living Southerner and bequeath them to every one yet to be born! May such sentiments be held universally in the outraged and down-trodden South, though in silence and stillness, until the now far-distant day shall arrive for just retribution for Yankee usurpation, oppression and atrocious outrages, and for deliverance and vengeance for the now ruined, subjugated and enslaved Southern States! . . . And now with my latest writing and utterance, and with what will be near my latest breath, I here repeat and would willing proclaim my unmitigated hatred to Yankee rule — to all political, social and business connections with Yankees, and the perfidious, malignant and vile Yankee race.

He then quickly and carefully prepared for the end and with a shotgun took his own life.[34]

The text printed here is that of the first edition published in Petersburg, Virginia, in 1832 by J. W. Campbell. In a few instances, typographical errors have been corrected and modern usage in punctuation has been adopted.

The footnotes that appear in this edition are contributed by both author and editor. Originally the author used asterisks, daggers, and double daggers in footnote citations. These author's notes are now indicated by arabic numerals numbered in sequence with the editor's notes, but they may be distinguished by the fact that the editor's notes are enclosed in square brackets. The editor has occasionally made additions to the author's notes, and these too are bracketed.

I am indebted to Dr. Samuel B. Knight of the Department of Chemistry at the University of North Carolina for valuable assistance in the interpretation of technical passages in the text.

J. CARLYLE SITTERSON

Chapel Hill, North Carolina
August 1960

---

[34] Edmund Ruffin, Jr., gave the details of his father's death in a letter to his sons, June 20, 1865; see *Tyler's Historical and Genealogical Magazine*, V (1924), 193–195.

1794   Born January 5, in Prince George County, Virginia, son of George Ruffin and Jane Lucas Ruffin.

1810   Attended briefly William and Mary College at Williamsburg, Virginia.

1812   Served as private in War of 1812, August 1812 to February 1813.

1813   Married Susan Travis of Williamsburg, Virginia, who died in 1846.

1813   Began active direction of plantation at Coggin's Point in Prince George County, Virginia.

1821   Published results of his experiments with marl in *American Farmer*, III (1821), 313–320.

1823   Elected to Virginia State Senate.

1832   Published his *Essay on Calcareous Manures* (Petersburg).

1833–1842   Published and edited *Farmers' Register*.

1841–1842   Published and edited for six months the monthly *Bank Reformer*.

1843   Served as Agricultural Surveyor for South Carolina, January to December; published *Report of the Commencement and Progress of the Agricultural Survey of South Carolina for 1843* (Columbia, S. C.).

1844   Moved residence to new plantation, Marlbourne, Hanover County, Virginia.

1852   Published *The Influence of Slavery, or Its Absence, on Manners, Morals, and Intellect* (N.P.).

1852   Elected president of Virginia Agricultural Society.

1853   Published *Premium Essay on Agricultural Education* (Richmond).

1855   Published *Essays and Notes on Agriculture* (Richmond).

1857   Published *Consequences of Abolition Agitation* (Washington); *Communications on Drainage and Other Connected Agricultural Subjects* (N.P.); *The Political Economy of Slavery* (Washington).

1859   Published *African Colonization Unveiled* (Washington); *Slavery and Free Labor Described and Compared* (N.P.); and *Notes on the Pine Trees of Lower Virginia and North Carolina* (N.P.).

1860   Published *Anticipations of the Future* (Richmond).

1860– Visited South Carolina, Georgia, and North Carolina to urge
1861   secession.
1861   Published *Agricultural, Geological and Descriptive Sketches of
       Lower North Carolina* . . . (Raleigh).
1861   Fired first gun at Fort Sumter, April 12.
1861– Lived during war at Beechwood, home of his son, Edmund, Jr.,
1865   in Prince George County, at Marlbourne in Hanover County,
       and at a farm in Amelia County, Virginia.
1865   Died in Amelia County, Virginia, June 17; buried at Marl-
       bourne, June 19.

AN

ESSAY

ON

CALCAREOUS MANURES.

BY EDMUND RUFFIN.

PETERSBURG, (Va.)
PUBLISHED BY J. W. CAMPBELL.
1832.
[facsimile of original title page]

PREFACE

THE object of this Essay is to investigate the peculiar features and qualities of the soils of our tide-water district, to show the causes of their general unproductiveness, and to point out means as yet but little used, for their effectual and profitable improvement. My observations are particularly addressed to the cultivators of that part of Virginia which lies between the sea coast and the falls of the rivers, and are generally intended to be applied only within those limits. By thus confining the application of the opinions which will be maintained, it is not intended to deny the propriety of their being further extended. On the contrary, I do not doubt but that they may correctly apply to all similar soils, under similar circumstances; for the operations of nature are conducted by uniform laws, and like causes must every where produce like effects. But as I shall rely for proofs on such facts as are either sufficiently well known already, or may easily be tested by any inquirer, I do not choose to extend my ground so far, as to be opposed by the assertion of other facts, the truth of which can neither be established nor overthrown by any available or sufficient testimony.

The peculiar qualities of our soils have been little noticed, and the causes of those peculiarities have never been sought — and though new and valuable truths may await the first explorers of this opening for agricultural research, yet they can scarcely avoid mistakes sufficiently numerous to moderate the triumph of success. I am not blind to the difficulties of the investigation, nor to my own unfitness to overcome them — nor should I have hazarded the attempt, but for the belief that such an investigation is all important for the improvement of our soil and agriculture, and that it was in vain to hope that it would be undertaken by those who were better qualified to do justice to the subject. I ask a deliberate hearing, and a strict scrutiny of my opinions, from those most interested in their truth. If a change in most of our lands, from hopeless sterility to a high state of productiveness, is a vain fancy, it will be easy to discover and expose the fallacy of my views: but if these views are well founded, none better deserve the attention of farmers, and nothing can more seriously affect the future agricultural prosperity of our country. No where ought such improvements to be more highly valued, or more eagerly sought, than among us, where so many causes have concurred to reduce our products, and the prices of our lands, to the lowest state, and are yearly extending want, and its consequence, ignorance, among the cultivators and proprietors.

In pursuing this inquiry, it will be necessary to show the truth of various facts and opinions, which as yet are unsupported by authority, and most of which have scarcely been noticed by agricultural writers, except to be denied. The number of proofs that will be required, and the discursive course through which they must be reached, may probably render more obscure the reasoning of an unpractised writer. Treatises on agriculture ought to be so written as to be clearly understood, though it should be at the expense of some other requisites of good writing — and in this respect, I shall be satisfied if I succeed in making my opinions intelligible to every reader, though many might well dispense with such particular explanations. Agricultural works are seldom considered as requiring very close attention; and therefore, to be made useful, they should be put in a shape suited to cursory and irregular reading. A truth may be clearly established — but if its important consequences cannot be regularly deduced for many pages afterwards, the premises will then probably have been forgotten, so that a very particular reference to them may be required. These considerations must serve as my apology for some repetitions — and for minute explanations and details, which some readers may deem unnecessary.

The theoretical opinions supported in this essay, together with my earliest experiments with calcareous manures, were published in the *American Farmer*, (III, 313,) in 1821. No reason has since induced me to retract any of the important positions then assumed. But the many imperfections in that publication, which grew out of my want of experience, made it my duty, at some future time, to correct its errors, and supply the deficiencies of proof, from the fruits of subsequent practice and observation. With these views, this essay was commenced and finished in 1826. But the work had so grown on my hands, that instead of being of a size suitable for insertion in an agricultural journal, it would have filled a volume. The unwillingness to assume so conspicuous a position, as the publication in that form would have required, and the fear that my work would be more likely to meet with neglect or censure than applause, induced me to lay it aside, and to give up all intention of publication. Since that time, the use of fossil shells as a manure has greatly increased, in my own neighbourhood and elsewhere, and it has been attended generally with all the improvement and profit that was expected. But from paying no regard to the theory of the operation of this manure, and not even taking warning from the known errors and losses of myself as well as others, most persons have used it injudiciously, and have damaged more or less of their lands. So many disasters of this kind, seemed likely to restrain the use of this valuable manure, and even to destroy its reputation, just as it was beginning rapidly to extend. This additional consideration has at last induced me to risk the publication of this essay. The experience of five more years, since it was written, has not contradicted any of the opinions then advanced — and no change has been made in the work, except in form, and by continuing the reports of experiments to the present time.

It should be remembered, that my attempt to convey instruction is confined to a single means of improving our lands, and increasing our profits: and though many other operations are, from necessity, incidentally noticed, my opinions or practices on such subjects are not referred to, as rules for good husbandry. In using calcareous manure for the improvement of poor soils, my labours have been highly successful — but that success is not necessarily accompanied by general good management and economy. To those who know me intimately, it would be unnecessary to confess the small pretensions that I have to the character of a good farmer — but to others, it may be required to explain why other improvements and practices of good husbandry have not been more aided by, and kept pace with, the effects of my use of calcareous manures.

E. R.

Coggin's Point, Virginia, }
  January 20*th*, 1832. }

CHAPTER I

GENERAL DESCRIPTION OF EARTHS AND SOILS

IT is very necessary that we should correctly distinguish *earths* and *soils*, and their many varieties: yet these terms are continually misapplied — and even among authors of high authority, no two agree in their definitions, or modes of classification. Where such differences exist, and no one known method is so free from material imperfections, as to be referred to as a common standard, it becomes necessary for every one who treats of soils, to define for himself — though perhaps he is thereby adding to the general mass of confusion already existing. This necessity must be my apology for whatever is new or unauthorized in the following definitions.

The *earths* important to agriculture, and which form nearly the whole of the known globe, are only three — *silicious, aluminous*, and *calcareous*.

Silicious earth, in its state of absolute purity, forms rock crystal. The whitest and purest sand may be considered as silicious earth in agriculture, though none is presented by nature entirely free from other ingredients. It is composed of very hard particles, not soluble in any common acid, and which cannot be made coherent by mixing with water. Any degree of coherence, or any shade of color that sand may exhibit, is owing to the presence of other substances. The solidity of the particles of sand renders them impenetrable to water, which passes between them as through a sieve. The hardness of its particles, and their loose arrangement, make sand incapable of absorbing moisture from the atmosphere, or of retaining any valuable vapour or fluid, with which it may have been in any manner supplied. Silicious earth is also quickly heated by the sun, which adds to the rapidity with which it loses moisture.

Aluminous or *argillaceous earth*, when dry, adheres to the tongue, absorbs water rapidly and abundantly, and when wet, forms a tough paste, smooth and soapy to the touch. By burning it becomes as hard as stone. Clays derive their adhesiveness from their proportion of aluminous earth. This also is white when pure, but is generally coloured deeply and variously — red, yellow, or blue — by metallic substances. When drying, aluminous earth shrinks greatly — it becomes a mass of very hard lumps, of various sizes, separated by cracks and fissures, which become so many little reservoirs of standing water, when filled by rains, and remain so, until the lumps, by slowly imbibing the water, are distended enough to fill the space occupied before.

Calcareous earth, or *carbonate of lime*, is *lime* combined with *carbonic acid*, and may be converted into pure or quick-lime by heat — and quick-lime, by exposure to the air, soon returns to its former state of calcareous earth. It forms marble, limestone, chalk, and shells, with very small admixtures of other substances. Thus the term *calcareous earth* will not be used by me to include either *lime* in its pure state, or any of the numerous combinations which lime forms with the various acids, except that alone which is beyond comparison the most abundant throughout the world, and most important as an ingredient of soils. Pure lime attracts all acids so powerfully, that it is never presented by nature except in combination with some one of them, and generally with the carbonic acid. When this compound is thrown into any stronger acid, as muriatic, nitric, or even strong vinegar — the lime being more powerfully attracted, unites with, and is dissolved by the stronger acid, and lets go the carbonic, which escapes with effervescence in the form of air. In this manner, the carbonate of lime, or calcareous earth, may easily be distinguished from silicious, and aluminous earth, and also from all other combinations of lime. [Appendix. A.]

Calcareous earth in its different forms has been supposed to compose as much as one-eighth part of the crust of the globe.[1] Very extensive plains in France and England are of chalk, pure enough to be nearly barren, and to prove that pure calcareous earth would be entirely so. No chalk is to be found in our country — and it is only from European authors that we can know any thing of its agricultural characters, when nearly pure, or when forming a very large proportion of the surface of the land. The whiteness of chalk repels the rays of the sun, while its loose particles permit water to pass through, as easily as sand: [2] and thus calcareous earth is remarkable for some of the worst qualities of both the other earths, and which it serves to cure in them (as I shall hereafter show) when used as a manure.

Most of those who have applied chemistry to agriculture, consider magnesia as one of the important earths.[3] Magnesia, like lime, is never found pure, but always combined with some acid, and its most general form is the carbonate of magnesia. But even in this, its usual and natural state, it exists in such very small quantities in soils, and is found so rarely, that its name seems a useless addition to the list of the earths of agriculture. For all practical purposes, gypsum (though only another combination of lime,) would more properly be arranged as a distinct earth, or elements of soils, as it is found in far greater abundance and purity, and certainly affects some soils and plants in a far more important manner than has yet been attributed to magnesia, in its natural form.

All the earths, when as pure as they are ever furnished by nature, are entirely barren, as might be inferred from the description of their qualities: nor would any addition of putres-

---

[1] [Parker Cleaveland, *An Elementary Treatise on Mineralogy and Geology* (Boston, 1816).]

[2] ["Terres" in *Nouveau Cours Complet d'Agriculture. . .* , ed. l'Abbé François Rozier, 13 vols. (Paris, 1785–1809).]

[3] [Davy, *Agricultural Chemistry* (Philadelphia, 1821'), p. 110. Where Ruffin cites *Agricultural Chemistry* without author he refers to the Philadelphia edition of Sir Humphrey Davy, *Elements of Agricultural Chemistry* (London, 1813).]

cent manures [4] enable either of the earths to support vegetable life.

The mixture of the three earths in due proportions, will correct the defects of all, and with a sufficiency of animal or vegetable matter, putrescent, and soluble in water, a *soil* is formed in which plants can extend their roots freely, yet be firmly supported, and derive all their needful supplies of air, water, and warmth, without being oppressed by too much of either. Such is the natural surface of almost all the habitable world: and though the qualities and value of soils are as various as the proportions of their ingredients are innumerable, yet they are mostly so constituted, that no one earthy ingredient is so abundant, but that the texture [5] of the soil is mechanically suited to some one valuable crop — some plants requiring a degree of closeness, and others of openness in the soil, which would cause other plants to decline or perish.

Soil seldom extends more than a few inches below the surface, as on the surface only are received those natural supplies of vegetable and animal matters, which are necessary to constitute soil. Valleys subject to inundation have soils brought from higher lands, and deposited by the water, and therefore are of much greater depth. Below the soil is the *subsoil*, which is also a mixture of two or more earths, but is as barren as the unmixed earths, because it contains very little putrescent matter, the only food for plants.

The qualities and value of soils depend on the proportions of their ingredients. We can easily comprehend in what manner silicious and aluminous earths, by their mixture, serve to cure the defects of each other — the open, loose, thirsty, and hot nature of sand being corrected by, and correcting in turn,

[4] *Putrescent* or *enriching manures*, are those formed of vegetable and animal matters, capable of putrefying, and thereby furnishing soluble food to plants. Farm-yard and stable manure, and the weeds and other growth of the fields left to die and rot on them, are almost the only enriching manures that we have used as yet.

[5] The *texture* of a soil means the disposition of its parts, which produce such sensible qualities, as being close, adhesive, open, friable, &c.

the close, adhesive, and water holding qualities of aluminous earth. This curative operation is merely mechanical — and in that manner it seems likely that calcareous earth, when in large proportions, also acts, and aids the corrective powers of both the other earths. This however is only supposition, as I have met with scarcely any such natural soil.

But besides the mechanical effects of calcareous earth, (which perhaps are weaker than those of the other two,) that earth has chemical powers far more effectual in altering the texture of soils, and for which a comparatively small quantity is amply sufficient. The chemical action of calcareous earth as an ingredient of soils, will be fully treated of hereafter: it is only mentioned in this place to avoid the apparent contradiction which might be inferred, if, in a general description of calcareous earth, I had omitted all allusion to qualities that will afterwards be brought forward as all important.

It seems most proper to class and name soils according to their *predominant* earthy ingredients, by which term, I mean those ingredients which exert the greatest power, and most strongly mark the character of the soil. The predominant ingredient (in this sense,) is not always the most abundant, and frequently is the least. If the most abundant was considered the predominant ingredient, and gave its name to the soil,[6] then almost every one should be called silicious, as that earth is seldom equalled in quantity by all the others united. If the earthy parts of a soil were two-thirds silicious, and one-third of aluminous earth, the peculiar qualities of the smaller ingredient would predominate over the opposing qualities of the sand, and the mixture would be a tenacious clay. If the same soil had contained only one-twentieth part of calcareous earth, that ingredient would have had more marked effects on the soil, than could have been produced by either doubling, or diminishing to half their quantity, the silicious and alumi-

[6] Which is the plan of the nomenclature of soils proposed by Rozier — See article "Terres."

nous earths, which formed the great bulk of the soil. If soils
were named according to certain proportions of their ingredients, (as proposed by Davy,[7]) a correct, though limited analysis of a soil would be required, before its name or character
could be given — and even then the name and character would
often disagree. But every farmer can know what are the most
marked good or bad qualities of his soils, as shown under
tillage, and those qualities can be easily traced to their predominant ingredients. By compounding a few terms, various
shades of difference may be designated with sufficient precision. A few examples will be sufficient to show how all may
be applied: —

A *silicious* or *sandy soil* has such a proportion of silicious
earth as to show more of its peculiar properties than those of
any other ingredient. It would be more or less objectionable
for its looseness, heat, and want of power to retain either
moisture or putrescent manure — and not for toughness, lia-
bility to become hard after wet ploughing, or any other quality of aluminous earth.

In like manner, an *aluminous* or *clayey soil*, would show
most strongly the faults of aluminous earth, though more than
half its bulk might be of silicious.

The term *loam* is not essential to this plan, but it is convenient, as it will prevent the necessity of frequent compounds
of other terms. It will be used for all soils formed with such
proportions of sand and aluminous earth, as not to be light
enough to be called sandy, nor stiff enough for clay soil.
Sandy loam and *clayey loam* would express its two extremes
— and *loamy sand* would be still lighter than the former, and
loamy clay stiffer than the latter.

In all compound names of soils, the last term should be considered as expressing the predominant earthy ingredient. Thus,
a *sandy loamy calcareous soil*, would be nearer to loam than
sand, and more marked by its calcareous ingredient than either.

[7] [*Agr. Chem.*, p. 139.]

Other ingredients of soils besides the earths, or any accidental or rare quality affecting their character considerably, may be described with sufficient accuracy by such additional terms as these — a *ferruginous gravelly silicious loam* — or a *vegetable calcareous clay*. [Appendix. B.]

CHAPTER II

ON THE SOILS AND STATE OF AGRICULTURE
OF THE TIDEWATER DISTRICT OF VIRGINIA

——————— "During several days of our journey, no spot was seen that was not covered with a luxuriant growth of large and beautiful forest trees, except where they had been destroyed by the natives for the purpose of cultivation. The least fertile of their pasture lands, without seeding, are soon covered with grass several feet in height; and unless prevented by cultivation, a second growth of trees rapidly springs up, which, without care or attention, attain their giant size in half the time that would be expected on the best lands in England." ———————

If the foregoing description was met with in a "Journey through Hindoostan," or some equally unknown region, no European reader would doubt but such lands were fertile in the highest degree — and even many of ourselves would receive the same impression. Yet it is no exaggerated account of the poorest natural soils in our own poor country, which are as remarkable for their producing luxuriant growths of pines, and broom grass, as for their improductiveness in every cultivated or valuable crop. We are so accustomed to these facts, that we scarcely think of their singularity, nor of the impropriety of calling any land barren, which will produce a rapid growth of any one plant. Indeed, by the rapidity of that growth, (or the fitness of the soil for its production,) we have in some measure formed a standard of the poverty of the soil.

With some exceptions to every general character, the tidewater district of Virginia may be described as generally level, sandy, poor, and free from any fixed rock, or any other than stones apparently rounded by the attrition of water. On much the greater part of the lands, no stone of any kind is to be found, of larger size than gravel. Pines of different kinds form the greater part of a heavy cover to the silicious soils in their virgin state, and mix considerably with oaks, and other growth of clay land. Both these kinds of soil, after being exhausted of their little fertility by cultivation, and "turned out" to recruit, are soon covered by young pines, which grow with vigour and luxuriance. This general description applies more particularly to the *ridges* which separate the *slopes* on different streams. The ridge lands are always level, and very poor — sometimes clayey, more generally sandy, but stiffer than would be inferred from the proportion of silicious earth they contain, which is caused by the fineness of its particles. Whortleberry bushes, as well as pines, are abundant on ridge lands — and numerous shallow basins are found, which are ponds of rain water in winter, but dry in summer. None of this large proportion of our lands, has paid the expense of clearing and cultivation, and much the greater part still remains under its native growth. Enough however has been cleared and cultivated in every neighbourhood, to prove its utter worthlessness, under common management. The soils of ridge lands vary between sandy loam, and clayey loam. It is difficult to estimate their general product under cultivation; but judging from my own experience of such soils, the product may be from five bushels of corn, or as much of wheat, to the acre, on the most clayey soils, to twelve bushels of corn on the most sandy — which would probably not yield three bushels of wheat, if it was there attempted.

The *slopes* extend from the ridges to the streams, or to the alluvial bottoms, and include the whole interval between neighbouring branches of the same stream. This class of soils forms another great body of lands; of a higher grade of fertility, though far from valuable. It is generally more sandy than the poorer ridge land, and when long cultivated, is more or less deprived of its soil, by the washing of rains, on every slight declivity. The washing away of three or four inches in depth, exposes a steril subsoil (or forms a "gall") which continues thenceforth bare of all vegetation: a greater declivity of the surface forms gullies several feet in depth, the earth from which, covers and injures the adjacent lower land. Most of this kind of land has been cleared, and greatly exhausted. Its virgin growth is often more of oak, hickory, and dogwood, than pine — but when turned out of cultivation, an unmixed growth of pine follows. Land of this kind in general has very little durability; its usual best product of corn may be for a few crops, eighteen or twenty bushels — and even as much as twenty-five bushels, from the highest grade. Wheat is seldom a productive or profitable crop on the slopes, the soil being generally too sandy. When such soils as these are called rich or valuable (as most persons would describe them,) those terms must be considered as only comparative — and such an application of them proves that truly fertile and valuable soils, are very scarce in Lower Virginia.

The only very rich and durable soils below the falls of our rivers, are narrow strips of highland along their banks, and the lowlands formed by the alluvion of the numerous smaller streams which water our country. These alluvial bottoms, though highly productive, are lessened in value by being generally too sandy, and by the damage they suffer from being often inundated by floods of rain. The best highland soils seldom extend more than half a mile from the river's edge — sometimes not fifty yards. These irregular margins are composed of loams of various qualities, but all highly valuable; and the best soils are scarcely to be surpassed, in their original fertility, and durability under severe tillage. Their nature and peculiarities will be again adverted to, and more fully described hereafter.

The simple statement of the general course of tillage to which our part of the country has been subjected, is sufficient to prove that great impoverishment of the soil has been the inevitable consequence. The small portion of rich river margins, was soon all cleared, and was tilled without cessation for many years. The clearing of the slopes was next commenced, and is not yet entirely completed. On these soils, the succession of crops was less rapid, or, from necessity, tillage was sooner suspended. If not rich enough for tobacco when first cleared, (or as soon as it ceased to be so,) land of this kind was planted in corn two or three years in succession, and afterwards every second year. The intermediate year between the crops of corn, the field was "rested" under a crop of wheat, if it would produce four or five bushels to the acre. If the sandiness, or exhausted condition of the soil, denied even this small product of wheat, that crop was probably not attempted — and instead of it, the field was exposed to close grazing, from the time of gathering one crop of corn, to that of preparing to plant another. No manure was applied, except on the tobacco lots; and this rotation of a grain crop every year, and afterwards every second year, was kept up as long as the field would produce five bushels of corn to the acre. When reduced below that product, and to less than the necessary expense of cultivation, the land was turned out to recover under a new growth of pines. After twenty or thirty years, according to the convenience of the owner, the same land would be again cleared, and put under similar scourging tillage, which however would then much sooner end, as before, in exhaustion. Such a general system is not yet every where abandoned — and many years have not passed, since such was the usual course on almost every farm.

How much our country has been impoverished during the last fifty years, cannot be determined by any satisfactory testimony. But however we may differ on this head, there are but few who will not concur in the opinion, that our system of cultivation has been every year lessening the productive power of our lands in general — and that no one county, no neigh-

bourhood, and but few particular farms, have been at all enriched, since their first settlement and cultivation. Yet many
of our farming operations have been much improved within
the last fifteen or twenty years. Driven by necessity, proprietors direct more personal attention to their farms — better
implements of husbandry are used — every process is more
perfectly performed — and whether well or ill directed, a
spirit of inquiry and enterprise has been awakened, which
before had no existence.

Throughout the country below the Falls, and perhaps thirty
miles above, if the best land be excluded, say one-tenth, the
remaining nine-tenths will not yield an average product of
ten bushels of corn to the acre; though that grain is best suited
to our soils in general, and far exceeds in quantity all other
kinds raised. Of course, the product of a large proportion of
the land, would fall below this average. Such crops, in very
many cases, cannot remunerate the cultivator. If our remaining woodland could be at once brought into cultivation, the
gross product of the country would be greatly increased, but
the *nett* product very probably diminished — as the general
poverty of these lands would cause more expense than profit
to accompany their cultivation under the usual system. Yet
every year we are using all our exertions to clear woodland,
and in fact seldom increase either nett or gross products — because nearly as much old exhausted land is turned out, as is
substituted by the newly cleared. Sound calculations of profit
and loss, would induce us to reduce the extent of our present
cultivation, by turning out every acre that yields less than the
total cost of its tillage.

No political truth is better established than that the population of every country will increase, or diminish, according to
its regular supply of food. We know from the census of 1830,
compared with those of 1820 and 1810, that our population is
nearly stationary, and in some counties, is actually lessening;
and therefore it is certain, that our agriculture is not increas-

ing the amount of food, or the means of purchasing food —
with all the assistance of the new land annually brought into
culture. A surplus population, with its deplorable consequences, is only prevented by the great current of emigration
which is continually flowing westward. No matter who emi-
grates, or with what motive — the enterprising or wealthy
citizen who leaves us to seek richer lands and greater profits,
and the slave sold and carried away on account of his owner's
poverty, concur in producing the same result, though with
very different degrees of benefit to those who remain. If this
great and continued drain from our population was stopped,
and our agriculture was not improved, want and misery would
work to produce the same results. Births would diminish, and
deaths would increase — and hunger and disease would keep
down population to that number, that the average products of
our agricultural and other labour could feed, and supply with
other means of living.

A stranger to our situation and habits might well oppose to
my statements the very reasonable objection, that no man
would, or could, long pursue a system of cultivation of which
the returns fell short of his expenses, including rent of land,
hire of labour, interest on the necessary capital, &c. Very true
— if he had to pay those expenses out of his profits, he would
soon be driven from his farm to a jail. But we own our land,
our labourers, and stock — and though the calculation of nett
profit, or of loss, is precisely the same, yet we are not ruined
by making only two per cent. on our capital, provided we
can manage to live on that income. If we live on still less, we
are actually growing richer (by laying up a part of our two
per cent.,) notwithstanding the most clearly proved regular
loss on our farming.

Our condition has been so gradually growing worse, that
we are either not aware of the extent of the evil, or are in a
great measure reconciled by custom to profitless labour. No
hope for a better state of things can be entertained, until we

shake off this apathy — this excess of contentment which makes no effort to avoid existing evils. I have endeavoured to expose what is worst in our situation as farmers — if it should have the effect of rousing any of my countrymen to the absolute necessity of some improvement, to avoid ultimate ruin, I hope also to point out to some of their number, if not to all, that the means for certain and highly profitable improvements, are completely within their reach. [Appendix. C.]

CHAPTER III

THE DIFFERENT CAPACITIES OF SOILS
FOR IMPROVEMENT

As far as the nature of the subjects permitted, the foregoing chapters have been merely explanatory and descriptive. The same subjects will be resumed and more fully treated in the course of the following argument, the premises of which, are the facts and circumstances that have been detailed. What I wish to prove will be stated in a series of propositions, which will now be presented at one view, and afterwards be separately discussed in their proper order.

Proposition 1. Soils naturally poor, and rich soils reduced to poverty by cultivation, are essentially different in their powers of retaining putrescent manures: and under like circumstances, the fitness of any soil to be enriched by these manures, is in proportion to what was its natural fertility.

2. The natural sterility of the soils of Lower Virginia is caused by such soils being destitute of calcareous earth, and their being injured by the presence and effects of vegetable acid.

3. The fertilizing effects of calcareous earth are chiefly produced by its power of neutralizing acids, and of combining putrescent manures with soils, between which there would otherwise be but little if any chemical attraction.[1]

---

[1] When any substance is mentioned as *combining* with one or more other substances, as different manures with each other, or with soil, I mean that a union is formed by chemical attraction, and not by simple mixture. *Mixtures* are made by mechanical means, and may be separated in like manner; but *combinations* are chemical, and require some stronger chemical attraction, to take away either of the bodies so united.

When two substances combine, they both lose their previous peculiar qualities, or *neutralize* them for each other, and form a third substance different from both. Thus, if certain known proportions of muriatic acid,

4. Poor and acid soils cannot be improved durably, or profitably, by putrescent manures, without previously making them calcareous, and thereby correcting the defect in their constitution.

5. Calcareous manures will give to our worst soils a power of retaining putrescent manures, equal to that of the best — and will cause more productiveness, and yield more profit, than any other improvement practicable in Lower Virginia.

Dismissing from consideration, for the present, all the others, I shall proceed to maintain the *First Proposition*.

Soils naturally poor, and rich soils reduced to poverty by cultivation, are essentially different in their power of retaining putrescent manures: and under like circumstances, the fitness of any soil to be enriched by these manures, is in proportion to what was its natural fertility.

The *natural fertility* of a soil is not intended to be estimated by the amount of its earliest product, when first brought under cultivation, because several temporary causes then operate either to keep down, or to augment the product. If land be cultivated immediately after the trees are cut down, the crop is greatly lessened by the numerous living roots, and consequent bad tillage — the excess of unrotted vegetable matter, and the coldness of the soil, from which the rays of the sun had been so long excluded. On the other hand, if cultivation is delayed one or two years, the leaves and other vegetable matters are rotted, and in the best state to supply food to plants, and are so abundant, that a far better crop will be raised than could have been obtained before, or perhaps will be again, without manure. For these reasons, the degree of natural fertility of any soil should be measured by its prod-

---

and pure or caustic soda, be brought together, their strong attraction will cause them to combine immediately. The violent corrosive acid quality of the one, and the equally peculiar alkaline taste and powers of the other, will *neutralize*, or entirely destroy each other — and the compound formed is common salt, the qualities of which are as strongly marked, but totally different from those of either of its constituent parts.

ucts after these temporary causes have ceased to act, which will generally take place before the third or fourth crop is gathered. According, then, to this definition, a certain degree of permanency in its early productiveness is necessary, to entitle a soil to be termed *naturally fertile*. It is in this sense, that I deny to any poor lands, except such as were naturally fertile, the capacity of being made rich by putrescent manures.

The foregoing proposition would by many persons be so readily admitted as true, that attempting to prove it would be deemed entirely superfluous. But many others will as strongly deny its truth, and can support their opposition by high agricultural authorities.

General readers, who may have no connexion with farming, must have gathered from the incidental notices in various literary works, that some countries or districts that were noted for their uncommon fertility or barrenness, as far back as any accounts of them have been recorded, still retain the same general character, through every change of policy, government, and even of the race of inhabitants. They know that for some centuries at least, there has been no change in the strong contrast between the barrenness of Norway, Brandenburg, and the Highlands of Scotland, and the fertility of Lombardy and Valencia. Sicily, notwithstanding its government is calculated to discourage industry, and production of every profitable kind, still exhibits that fertility for which it was celebrated two thousand years ago. It seems a necessary inference from the many statements of which these are examples, that the labours of man have been but of little avail in altering permanently the characters and qualities given to soils by nature.

Most of our experienced practical cultivators, through a different course, have arrived at the same conclusion. Their practice has taught them the truth of this proposition — and the opinions thus formed have profitably directed their most important operations. They are accustomed to estimate the worth of land by its natural degree of fertility — and by the same rule they are directed on what soils to bestow their scanty stock of manure, and where to expect exhausted fields to recover by rest, and their own unassisted powers. But content with knowing the fact, this useful class of farmers have never inquired for its cause — and their opinions on this subject, as on most others, have not been communicated so as to benefit others.

But if all literary men who are not farmers, and all practical cultivators who seldom read, admitted the truth of my proposition, it would avail but little for improving our agricultural operations — and the only prospect of its being usefully disseminated, is through that class of farmers who have received their first opinions of improving soils, from books, and whose subsequent plans and practice have grown out of those opinions. If poor natural soils cannot be durably or profitably improved by putrescent manures, this truth should not only be known, but kept constantly in view, by every farmer who can hope to improve with success. Yet it is a remarkable fact, that the difference in the capacities of soils for receiving improvement, has not attracted the attention of scientific farmers — and the doctrine has no direct and positive support from the author of any treatise on agriculture, English or American, that I have been able to consult. On the contrary, it seems to be considered by all of them, that to collect and apply as much vegetable and animal manure as possible, is sufficient to ensure profit to every farmer, and fertility to every soil. They do not tell us that numerous exceptions to that rule will be found, and that many soils of apparent good texture, if not incapable of being enriched from the barn-yard, would at least cause more loss than clear profit, by being improved from that source.

When it is assumed that the silence of every distinguished author as to certain soils being incapable of being profitably enriched, amounts to ignorance of the fact, or a tacit denial of its truth — it may be objected that the exception was not omitted from either of these causes, but because it was established and undoubted. This is barely possible: but even if such was the case, their silence has had all the ill consequences that could have grown out of a positive denial of any exceptions to the propriety of manuring poor soils. Every zealous young farmer, who draws most of his knowledge and opinions from books, adopts precisely the same idea of their directions — and if he owns barren soils, he probably throws away his labour and manure for their improvement, for years, before experience compels him to abandon his hopes, and acknowledge that his guides have led him only to failure and loss. Such farmers as I allude to, by their enthusiasm and spirit of enterprise, are capable of rendering the most important benefits to agriculture. Whatever may be their impelling motives, the public derives nearly all the benefit of their successful plans — and their far more numerous misdirected labours, and consequent disappointments, are productive of national, still more than individual loss. The occurrence of only a few such mistakes, made by reading farmers, will serve to acquit men of combating a shadow — and there are few of us who cannot recollect some such examples.

But if the foregoing objection has any weight in justifying European authors in not naming this exception, it can have none for those of our country. If it is admitted that soils naturally poor are incapable of being enriched, with profit, that admission must cover three-fourths of all the highland in the tide-water district. Surely no one will contend that so sweeping an exception was silently understood by the author of *Arator*,[2] as qualifying his exhortations to improve our lands: and if no such exception was intended to be made, then will his directions for improvement, and his promises of reward,

[2] [John Taylor of Caroline County, Va. (1753–1824), political writer, agriculturist, and statesman, was the author of *Arator*, the most influential volume of essays on agriculture published in the ante-bellum South prior to Ruffin's work. See *D. A. B.*, XVIII, 331–333.]

be found equally fallacious, for the greater portion of the
country, to benefit which his work was specially intended.
The omission of any such exception by the writers of the
United States, is the more remarkable, as the land has been so
recently brought under cultivation, that the original degree
of fertility of almost every farm may be known to its owner,
and compared with the after progress of exhaustion or improvement.

I might quote many authorities to prove that I have correctly stated what is the fair and only inference to be drawn
from agricultural books, respecting the capacity of poor soils
to receive improvement. But a few of the most strongly
marked passages in *Arator* will be fully sufficient for this purpose. The venerated author of that work was too well acquainted with the writings of European agriculturists, to have
mistaken their doctrines in this important particular. A large
portion of his useful life was devoted to the successful improvement of exhausted, but originally fertile lands. His instructions for producing similar improvements are expressly
addressed to the cultivators of the eastern parts of Virginia
and North Carolina, and are given as applicable to all our
soils, without exception. Considering all these circumstances,
the conclusions which are evidently and unavoidably deduced
from his work, may be fairly considered, not only as supported by his own experience, but as concurring with the
general doctrine of improving poor soils, maintained by previous writers.

At page 54, third edition of *Arator*, *"inclosing"* [i. e. leaving
fields to receive their own vegetable cover, for their improvement during the years of rest,] is said to be "the most powerful means of fertilizing the earth" — and the process is declared to be rapid, the returns near, and the gain great.

Page 61. "If these few means of fertilizing the country
[cornstalks, straw, and animal dung,] were skilfully used, they
would of themselves suffice to change its state from sterility

to fruitfulness." — "By the litter of Indian corn, and of small grain, and of penning cattle, managed with only an inferior degree of skill, in union with inclosing, I will venture to affirm that a farm may in ten years be made to double its produce, and in twenty to quadruple it."

No opinions could be more strongly or unconditionally expressed than these. No reservation or exception is made. I may safely appeal to each of the many hundreds who attempted to obey these instructions, to declare whether any one considered his own naturally poor soils excluded from the benefit of these promises — or whether a tithe of that benefit was realized on any farm composed generally of such soils. In a field of mine that has been secured from grazing since 1814, and cultivated on the four shift rotation, the produce of a marked spot has been measured every fourth year (when in corn) since 1820. The difference of product has been such as the differences of season might have caused — and the last crop (in 1828) was worse than those of either of the two preceding rotations. There is no reason to believe that even the smallest increase of productive power has taken place.

It is far from my intention, by these remarks, to deny the propriety, or to question the highly beneficial results, of applying the system of improvement recommended by *Arator*, to soils originally fertile. On the contrary, it is as much my object to maintain the facility of restoring to worn lands their natural degree of fertility, by vegetable applications, as it is to deny the power of exceeding that degree, however low it may have been. One more quotation will be offered, because its recent date and the source whence it is derived, furnish the best proof that it is still the received opinion among agricultural writers, that all soils may be profitably improved, by putrescent manures. An article in the *American Farmer*, of October 14th 1831, on "manuring large farms," by the editor, contains the following expressions. "——— By proper exertions, every farm in the United States can be manured with less expense than the surplus profits arising from the manure would come to. This we sincerely believe, and we have arrived at this conclusion from long and attentive observation. We never yet saw a farm that we could not point to means of manuring, and bring into a state of high and profitable cultivation at an expense altogether inconsiderable when contrasted with the advantages to be derived from it." The remainder of the article shows that putrescent manures are principally relied on to produce these effects: marsh and swamp mud are the only kinds referred to that are not entirely putrescent in their action, and mud certainly cannot be used to manure every farm. Mr. Smith,[3] having been long the conductor of a valuable agricultural journal, as a matter of course, is extensively acquainted with the works and opinions of the best writers on agriculture; and therefore, his advancing the foregoing opinions, as certain and undoubted, is as much a proof of the general concurrence of preceding writers, as if the same had been given as a digest of their precepts.

Some persons will readily admit the great difference in the capacities of soils for improvement, but consider a deficiency of clay only to cause the want of power to retain manures. The general excess of sand in our poor lands might warrant this belief in a superficial and limited observer. But though clay soils are more rarely met with, they present, in proportion to their extent, full as much poor land. The most barren and worthless soils in the county of Prince George, are also the stiffest. A poor clay soil, will retain manure longer than a poor sandy soil — but it will not the less certainly lose its acquired fertility at a somewhat later period. When it is considered how much more manure is required by clay soils, it is doubtful whether the improvement of the sandy soil would not be attended with more profit — or more properly speaking, with less actual loss.

[3] [Gideon B. Smith edited the *American Farmer* (Baltimore) from Sept. 1830 to Oct. 1833.]

It is true that the capacity of a soil for improvement is greatly affected by its texture, shape of the surface, and its supply of moisture. Dry, level, clay soils, will retain manure longer, than if they were sandy, hilly, or wet. But however important these circumstances may be, neither the presence or absence of any of them can cause the differences of capacity for improvement. There are rich and valuable soils with one or more of all these faults — and there are soils the least capable of improvement, free from objection as to their texture, degree of moisture, or inclination of their surface. Indeed the great body of our poor ridge lands, are more free from faults of this kind, than soils of far greater productiveness usually are. Unless then some other, and far more powerful obstacle to improvement exists, why should not all our woodland be highly enriched, by the hundreds, or thousands of crops of leaves which have successively fallen and rotted there? Notwithstanding this vegetable manuring, which infinitely exceeds all that the industry and patience of man can possibly equal, most of our woodland remains poor — and this one fact (which at least is indisputable,) ought to satisfy all of the impossibility of enriching such soils by putrescent manures only. Some few acres may be highly improved, by receiving all the manure derived from the offal of the whole farm — and entire farms, in the neighbourhood of towns, may be kept rich by continually applying large quantities of purchased manures. But no where can a farm be found, which has been improved beyond its original fertility, by means of the vegetable resources of its own arable fields. If this opinion is erroneous, nothing is easier than to prove my mistake, by adducing undoubted examples of such improvements having been made.

But a few remarks will suffice on the capacity for improvement of worn lands, which were originally fertile. With regard to these soils, I have only to concur in the received opinion of their fitness for durable and profitable improve-

ment by putrescent manures. After being exhausted by cultivation, they will recover their productive power, by merely
being left to rest for a sufficient time, and receiving the manure made by nature, of the weeds and other plants that grow
and die upon the land. Even if robbed of the greater part of
that supply, by the grazing of animals, a still longer time will
serve to obtain the same result. The better a soil was at first,
the sooner it will recover by these means, or by artificial
manuring. On soils of this kind, the labours of the improving
farmer meet with success and reward — and whenever we
hear of remarkable improvements of poor land by putrescent
manures, further inquiry will show us that these poor lands
had once been rich.

The continued fertility of certain countries for hundreds
or even thousands of years, does not prove that the land could
not be, or had not been, exhausted by cultivation: but only
that it was slow to exhaust and rapid in recovering — so that
whatever repeated changes may have occurred in each particular tract, the whole country taken together always retains
a high degree of productiveness. Still the same rule will apply
to the richest and the poorest soils — that each exerts strongly
a force to retain as much fertility as nature gave them — and
that when worn and reduced, each may easily be restored to
its original state, but cannot be raised higher, with either
durability or profit.

CHAPTER IV

EFFECTS OF THE PRESENCE OF CALCAREOUS EARTH IN SOILS

PROPOSITION 2. *The natural sterility of the soils of Lower Virginia is caused by such soils being destitute of calcareous earth, and their being injured by the presence and effects of vegetable acid.*

The means which would appear the most likely to lead to the causes of the different capacities of soils for improvement, is to inquire whether any known ingredient or quality is always to be found belonging to improvable soils, and never to the unimprovable — or which always accompanies the latter, and never the former kind. If either of these results can be obtained, we will have good ground for supposing that we have discovered the general cause of fertility, in the one case — or of barrenness, in the other: and it will follow, that if we can supply to barren soils the deficient beneficial ingredient — or can destroy that which is injurious to them — that their incapacity for receiving improvement will be removed. All the common ingredients of soils, as sand, clay, or gravel — and such qualities as moisture or dryness — a level, or a hilly surface — however they may affect the value of soils, are each found exhibited in a remarkable degree, in both the fertile and the steril. The abundance of putrescent vegetable matter might well be considered the cause of fertility, by one who judged only from lands long under cultivation. But though vegetable matter in sufficient quantity is essential to the existence of fertility, yet will this substance also be found inadequate, as its cause. Vegetable matter abounds in all rich land, it is admitted;

but it has also been furnished by nature, in quantities exceeding all computation, to the most barren soils we own.

But there is one ingredient, of which not the smallest proportion can be found in any of our poor soils, and which, wherever found, indicates a soil remarkable for natural and durable fertility. This is *calcareous earth*. These facts alone, if sustained, will go far to prove that this earth is the cause of fertility, and the cure for barrenness.

On some part of most farms touching tide-water, either muscle or oyster shells are found mixed with the soil. Oyster shells are confined to the lands on salt water, where they are very abundant, and sometimes extend through large fields. Higher up the rivers, muscle shells only are to be seen thus deposited by nature, and they decrease as we approach the Falls. The proportion of shelly land in the countries highest on tide-water, is very small — but the small extent of these spots does not prevent, but rather aids, the investigation of the peculiar qualities of such soils. Spots of shelly land, not exceeding a few acres in extent, could not well have been cultivated differently from the balance of the fields, of which they formed parts — and therefore they can be better compared with the worse soils, under like treatment. Every acre of shelly land is, or has been, remarkable for its richness, and still more for its durability. There are few farmers among us who have not heard described tracts of shelly soil on Nansemond and York rivers, which are celebrated for their long resistance of the most exhausting system of tillage, and which still remain fertile, notwithstanding all the injury which they must have sustained from their severe treatment. We are told that on some of these lands, corn has been raised every successive year, without any help from manure, for a longer time than the owners could remember, or could be informed of, correctly. But without relying on any such remarkable cases, there can be no doubt but that every acre of our shelly land has been at least as much tilled, and as little manured, as any in the country — and that it is still the richest and most valuable of all our old cleared land.

The fertile but narrow strips along the banks of our rivers, (which form the small portion of our highland of first rate quality,) seldom extend far without exhibiting spots in which shells are visible, so that the eye alone is sufficient to prove the soil of such places to be calcareous. The similarity of natural growth, and of all other marks of character are such, that the observer might very naturally infer that former presence of shells had given the same valuable qualities to all these soils — but that they had so generally rotted, and been incorporated with the other earths, that they remained visible only in a few places, where they had been most abundant. The accuracy of this inference will hereafter be examined.

The natural growth of the shelly soils, (and of those adjacent of similar value,) is entirely different from that of the great body of our lands. Whatever tree thrives well on the one, is seldom found on the other class of soils — or if found, it shows plainly by its imperfect and stunted condition, on how unfriendly a soil it is placed. To the rich river margins are almost entirely confined the black or wild locust, hackberry or sugar nut tree, and papaw. The locust is with great difficulty eradicated, or the newer growths kept under on cultivated lands, and from the remarkable rapidity with which it springs up, and increases in size, it forms a serious obstacle to the cultivation of the river banks. Yet on the woodland only a mile or two from the river, not a locust is to be seen. On shelly soils, pines and broom grass cannot thrive, and are rarely able to maintain the most sickly growth.

Some may say that these striking differences of growth do not so much show a difference in the constitution of the soils, as in their state of fertility — or that one class of the plants above named delights in rich, and the other, in poor land. No plant prefers poor to rich soil — can thrive better on a scarcity of food, than with an abundant supply. Pine, broom grass,

and sorrel delight in a class of soils that are generally unproductive — but not on account of their poverty, for all these
plants show, by the greater or less vigour of their growth, the
abundance or scarcity of vegetable matter in the soil. But on
this class of soils, no quantity of vegetable manure could make
locusts flourish, though they will grow rapidly on a calcareous
hillside, from which all the soil capable of supporting other
plants, has been washed away.

In thus describing and distinguishing soils by their growth,
let me not be understood as extending those rules to other
soils and climates than our own. It is well established that
changes of kind in successive growths of timber have occurred in other places, without any known cause — and a
difference of climate will elsewhere produce effects, which
here would indicate a change of soil.

Some rare exceptions to the general fertility of shelly lands
are found where the proportion of calcareous earth is in great
excess. Too much of this ingredient causes even a greater degree of sterility than its total absence. This cause of barrenness is very common in France and England (on chalk soils,)
and very extensive tracts are not worth the expense of cultivation, or improvement. The few small spots that are rendered
barren here, are seldom (if ever) so affected by the excess of
oyster or muscle shells in the soil. These effects generally are
caused by beds of fossil sea shells, which in some places reach
the surface, and are thus exposed to the plough. These spots
are not often more than thirty feet across, and their nature is
generally evident to the eye; and if not, is so easily determined
by chemical tests, as to leave no reason for confounding the
injurious and beneficial effects of calcareous earth. This exception to the general fertilizing effect of this ingredient of
our soils, would scarcely require naming, but to mark what
might be deemed an apparent contradiction. But this exception, and its cause, must be kept in mind, and considered as
always understood and admitted, throughout all my remarks,

and which therefore it is not necessary to name specially, when the general qualities of calcareous earth are spoken of.

In the beginning of this chapter, I advanced the important fact that none of our poor soils contain naturally the least particle of calcareous earth. So far, this is supported merely by my assertion — and all those who have studied agriculture in books, will require strong proof before they can give credit to the existence of a fact, which is either unsupported, or indirectly denied, by all written authority. Others, who have not attended to such descriptions of soils in general, may be too ready to admit the truth of my assertion — because, not knowing the opinions on this subject, heretofore received and undoubted, they would not be aware of the importance of their admission.

It is true that no author has said expressly that every soil contains calcareous earth. Neither has any one stated that every soil contains some silicious, or aluminous earth. But the manner in which each has treated of soils and their constituent parts, would cause their readers to infer, that neither of these three earths is ever entirely wanting — or at least that the entire absence of the calcareous, is as rare as the absence of silicious or aluminous earth. Nor are we left to gather this opinion solely from indirect testimony, as the following examples, from the highest authorities, will prove. Davy says, "four earths generally abound in soils, the aluminous, the silicious, the calcareous, and the magnesian" [1] — and the soils of which he states the constituent parts, obtained by chemical analysis, as well as those reported by Kirwan, and by Young, all contain some proportion (and generally a large proportion) of calcareous earth.[2] Kirwan states the component parts of a soil which contained thirty-one per cent. of calcareous earth,

---

[1] Davy, *Agr. Chem.*, Lecture 1.

[2] *Ibid.*, Lecture 4 — Kirwan on Manures — and Young's Prize Essay on Manures. [Probably Richard Kirwan (1733–1812), Irish chemist, author of *The Manures . . . Applicable to the Various . . . Sorts of Soils . . .* (London, 1796); Arthur Young, *Essay on Manures* (London, 1804).]

and he supposes that proportion neither too little nor too much.[3] Young mentions soils of extraordinary fertility containing seventeen and twenty per cent., besides others with smaller proportions of calcareous earth — and says that Bergman found thirty per cent. in the best soil he examined.[4] Rozier speaks still more strongly for the general diffusion, and large proportions of this ingredient of soils. In his general description of earths and soils, he gives examples of the supposed composition of the three grades of soils which he designates by the terms *rich*, *good*, and *middling soils:* to the first class he assigns a proportion of one-tenth, to the second, one-fourth, and to the last, one-half of its amount, of calcareous earth. The fair interpretation of the passage is that the author considered these large proportions as general, in France — and he gives no intimation of any soil entirely without calcareous earth.[5]

American writers also suppose the general presence of this ingredient of soil: but their opinions on this subject are merely echos of European descriptions of soils. They seem neither to have suspected that so important a difference existed, nor to have made the least investigation by actual analysis, to sustain the false character thus given to the soils of our country. [Appendix. D.]

With my early impressions of the nature and composition of soils, derived from the general descriptions given in books,

[3] Kirwan on Manures, Article Clayey Loam.

[4] Young's *Essay on Manures.* [Young probably had reference to Torbern Olof Bergman (1734–1784), Swedish chemist and naturalist, many of whose works were translated into English.]

[5] "*Composition of soils.* Examples of the various composition of soils: *Rich soil*; silicious earth, 2 parts; aluminous, 6; calcareous, 1; vegetable earth, (*humus*) 1; in all, 10 parts. *Good soil*; silicious, 3 parts; aluminous 4; calcareous 2½; vegetable earth, ½ of 1 part; in all, 10 parts. *Middling soil (sol mediocre)*; silicious, 4 parts; aluminous, 1; calcareous, 5 parts, less by some atoms of vegetable earth; in all, 10 parts. We see that it is the largest proportion of aluminous earth, that constitutes the greatest excellence of soils; and we know that independently of their harmony of composition, they require a sufficiency of depth." — From the Article "*Terres*" in *Cours Complet d'Agriculture Pratique*, ed. L'Abbé Rozier.

it was with surprise, and some distrust, that when first attempting to analyze soils, in 1817, I found most specimens destitute of calcareous earth. The trials were repeated with care and accuracy, on soils from various places — until I felt authorized to assert without fear of contradiction, that no naturally poor soil, below the Falls, contains the smallest proportion of calcareous earth. Nor do I believe that any exception to this peculiarity of constitution can be found in any poor soil above the Falls: but though these are far more extensive and important in other respects, they are beyond the district, within the limits of which I propose to confine my investigation.

These results are highly important, whether considered merely as serving to establish my proposition, or as showing a radical difference between most of our soils, and those of the best cultivated parts of Europe. Putting aside my argument to establish a particular theory of improvement, the ascertained fact of the universal absence of calcareous earth in our poor soils leads to this conclusion: that profitable as calcareous manures have been found to be in countries where the soils are generally calcareous in some degree, they must be far more so on our soils that are quite destitute of that necessary earth.

CHAPTER V

RESULTS OF THE CHEMICAL EXAMINATIONS OF VARIOUS SOILS

PROPOSITION 2. *Continued.*

THE certainty of any results of chemical analysis would be doubted by most persons who have paid no attention to the means employed for such operations: and their incredulity will be the more excusable, when such results are reported by one knowing very little of the science of chemistry, and whose limited knowledge was gained without aid or instruction, and was sought solely with the view of pursuing this investigation. Appearing under such disadvantages, it is therefore the more incumbent on me to show my claim to accuracy, or so to explain my method, as to enable others to detect its errors, if any exist. To analyze a specimen of soil completely, requires a degree of scientific acquirement and practical skill, to which I make no pretension. But merely to ascertain the absence of calcareous earth — or if present, to find its quantity — requires but little skill, and less science.

The methods recommended by different agricultural chemists for ascertaining the proportion of calcareous earth in soils, agree in all material points. Their process will be described, and made as plain as possible. A specimen of soil of convenient size is dried, pounded, and weighed, and then thrown into muriatic acid, diluted with three or four times its quantity of water. The acid combines with, and dissolves the *lime* of the calcareous earth, and its other ingredient, the *carbonic acid*, being disengaged, rises through the liquid in the form of *gas*, or air, and escapes with effervescence. After the mixture has been well shaken, and has stood until all effervescence is over, (the fluid still being somewhat acid to the taste,) the whole is poured into a piece of blotting paper folded so as to fit within a glass funnel. The fluid containing the dissolved lime passes through the paper, leaving behind the clay and silicious sand, and any other solid matter; over which pure water is poured and passed off several times, so as to wash off all remains of the dissolved lime. These filtered washings are added to the solution, to all of which is then poured a solution of *carbonate of potash*. The two dissolved salts thus thrown together, (*muriate of lime*, composed of muriatic acid and lime — and *carbonate of potash*, composed of carbonic acid and potash,) immediately decompose each other, and form two new combinations. The muriatic acid leaves the lime, and combines with the potash, for which it has a stronger attraction — and the muriate of potash thus formed, being a soluble salt, remains dissolved and invisible in the water. The lime and carbonic acid being in contact, when let loose by their former partners, instantly unite, and form *carbonate of lime*, or calcareous earth, which being insoluble, falls to the bottom, is separated by filtering paper, is washed, dried, and weighed, and thus shows the proportion contained by the soil.[1]

In this process, the carbonic acid which first composed part of the calcareous earth, escapes into the air, and another supply is afterwards furnished from the decomposition of the carbonate of potash. But this change of one of its ingredients does not alter the quantity of the calcareous earth, which is always composed of certain invariable proportions of its two component parts; and when all the lime has been precipitated as above directed, it will necessarily be combined with precisely its first quantity of carbonic acid.

This operation is so simple, and the means for conducting it so easy to obtain, that it will generally be the most convenient mode for finding the proportion of calcareous earth in those

[1] More full directions for the analysis of soils may be found in Kirwan's *Essay on Manures*, Rozier's Dictionary, and especially in Davy's *Agricultural Chemistry*.

manures that are known to contain it abundantly, and where
an error of a few grains cannot be very material. But if a very
accurate result is necessary, this method will not serve, on account of several causes of error which always occur. Should
no calcareous earth be present in a soil thus analyzed, the
muriatic acid will take up a small quantity of aluminous earth,
which will be precipitated by the carbonate of potash, and
without further investigation, would be considered as so much
calcareous earth. If any compounds of lime and vegetable
acids are present, (which for reasons hereafter to be stated, I
believe to be not uncommon in soils,) some portion of them
may be dissolved, and appear in the result as *carbonate* of
lime, though not an atom of that substance was in the soil.
Thus, every soil examined by this method of precipitation,
will yield some small result of what would appear as calcareous
earth, though actually destitute of such an ingredient. The
inaccuracies of this method were no doubt known (though
passed over without notice) by Davy, and other men of science who have recommended its use: but as they considered
calcareous earth merely as one of the earthy ingredients of
soil, operating mechanically, (as do sand and clay,) on the
texture of the soil, they would scarcely suppose that a difference of a grain or two could materially affect the practical
value of an analysis, or the character of the soil under examination.[2]

The pneumatic apparatus proposed by Davy,[3] as another
means for knowing the proportion of calcareous earth in soils,
is liable to none of these objections; and when some other

[2] "Chalks, calcareous marls, or powdered limestone, *act merely by forming a useful earthy ingredient in the soil*, and their efficacy is proportioned
to the deficiency of calcareous matter, which in larger or smaller quantities
seems to be an essential ingredient of all fertile soils; necessary *perhaps* to
their proper texture, and as an ingredient in the organs of plants." (Davy's
Agr. Chem., page 21 — and further on he says) "Chalk and marl or carbonate of lime *only improve the texture of a soil, or its relation to absorption*;
it acts merely as one of its earthy ingredients."

[3] See the plate and description in Lecture fourth of *Agricultural Chemistry.*

causes of errors peculiar to this method, are known and guarded against, its accuracy is almost perfect, in ascertaining the quantity of calcareous earth — to which substance alone, its use is limited. The correctness of this mode of analysis depends on two well established facts in Chemistry — 1st. That the component parts of calcareous earth always bear the same proportion to each other — and these proportions are as forty-three parts (by weight) of carbonic acid, to fifty-seven of lime. 2nd. That the carbonic acid gas which two grains of calcareous earth will yield, is equal in bulk to one ounce of fresh water.[4] The process with this apparatus disengages, confines, and measures the gas evolved — and for every measure equal to the bulk of an ounce of water, the operator has only to allow two grains of calcareous earth in the soil acted on. It is evident that the result can indicate the presence of lime in no other combination except that which forms calcareous earth — nor any other earth, except carbonate of magnesia, which might be mistaken for calcareous earth, but which is too rare, and occurs in proportions too small, to cause any material error.

But if it is only desired to know whether calcareous earth is entirely wanting in any soil — or to test my assertion that so great a proportion of our soils are of that character — it may be done with far more ease than by either of the foregoing methods, and without apparatus of any kind. Let a handful of the soil (without drying or weighing) be thrown into a large drinking glass, containing enough of pure water to cover the soil about two inches. Stir it until all the lumps have disappeared, and the water has certainly taken the place of all the atmospheric air which the soil had inclosed. Remove any vegetable fibres, or froth, from the surface of the liquid, so as to have it clear. Then pour in gently about a tablespoonful of undiluted muriatic acid, which by its greater weight will sink, and penetrate the soil, without any agitation being neces-

4 *Agr. Chem.*

sary for that purpose. If any calcareous earth is present, it will quickly begin to combine with the acid, throwing off its carbonic acid in *gas*, which cannot fail to be observed as it escapes, as the gas from only eight grains, is equal in bulk to a gill measure. Indeed, the product of only a single grain of calcareous earth, would be abundantly plain to the eye of a careful operator, though it might be the whole amount of gas from two thousand grains of soil. If no effervescence is seen, even after adding more acid and gently stirring the mixture, then it is absolutely certain that the soil contained not the smallest portion of carbonate of lime — nor of the only other substance which might be mistaken for it, the carbonate of magnesia.

The examinations of all the soils that will be here mentioned, were made in the pneumatic apparatus, except some of those which evidently evolved no gas, and when no other result was required. As calcareous earth is plainly visible to the eye in all shelly soils, they only need examination to ascertain its proportion. A few examples will show what proportions we may find, and how greatly they vary, even in soils apparently of equal value.

1. Soil, a black clayey loam, from the top of the high knoll at the end of Coggin's Point, on James River, containing fragments of muscle shells throughout. Never manured, and supposed to have been under scourging cultivation and close grazing from the first settlement of the country: still capable of producing twenty-five or thirty bushels of corn — and the soil well suited to wheat. One thousand grains, cleared by a fine sieve of all coarse shelly matter, (as none can act on the soil until minutely divided,) yielded sixteen ounce measures of carbonic acid gas, which showed the finely divided calcareous earth to be thirty-two grains.

2. One thousand grains of similar soil from another part of the same field, treated in the same manner, gave twenty-four grains of finely divided calcareous earth.

3. From the east end of a small island, at the end of Coggin's Point, surrounded by the river, and tide marsh. Soil, dark brown loam, much lighter than the preceding specimens, though not sandy — under like exhausting cultivation — then capable of bringing thirty to thirty-five bushels of corn — not a good wheat soil, ten or twelve bushels being probably a full crop. One thousand grains yielded eight grains of coarse shelly matter, and eighty-two of finely divided calcareous earth.

4. From a small spot of sandy soil, almost bare of vegetation, and incapable of producing any grain, though in the midst of very rich land, and cleared but a few years. Some small fragments of fossil sea shells being visible, proved this barren spot to be calcareous, which induced its examination. Four hundred grains yielded eighty-seven of calcareous earth — nearly twenty-two per cent. This soil was afterwards dug and carried out as manure.

5. Black friable loam, from Indian Fields, on York River. The soil was a specimen of a field of considerable extent, mixed throughout with oyster shells. Though light and mellow, the soil did not appear to be sandy. Rich, durable, and long under exhausting cultivation.

1260 grains of soil yielded

168	of coarse shelly matter, separated mechanically,
8	finely divided calcareous earth.

The remaining solid matter, carefully separated, (by agitation and settling in water,) consisted of

130	grains of fine clay, black with putrescent matter, and which lost more than one-fourth of its weight by being exposed to a red heat,
875	white sand, moderately fine,
20	very fine sand,
36	lost in the process.
1061	

6. Oyster shell soil of the best quality from the farm of

Wills Cowper Esq. on Nansemond River — never manured, and supposed to have been cultivated in corn as often as three years in four, since the first settlement of the country — now yields (by actual measurement) thirty bushels of corn to the acre — but is very unproductive in wheat. A specimen taken from the surface to the depth of six inches, weighed altogether 242 dwt., which consisted of

126	of shells and their fragments, separated by the sieve,
116	remaining finely divided soil.

500 grains of the finely divided part, consisted of

18	carbonate of lime,
330	silicious sand — none very coarse,
94	impalpable aluminous and silicious earth,
35	putrescent vegetable matter — none coarse or un-rotted,
23	loss.
500	

It is unnecessary to cite any particular trials of our poor soils, as it has been stated that all are entirely destitute of calcareous earth, excluding the rare, but well marked exceptions of its great excess — of which an example has been given in the soil marked 4, in the foregoing examinations.

Unless then I am mistaken in supposing that these facts are universally true, the certain results of chemical analysis completely establish these general rules — viz.

That all calcareous soils are naturally fertile and durable in a very high degree — and

That all soils naturally poor are entirely destitute of calcareous earth.

It then can scarcely be denied that calcareous earth must be the cause of the fertility of the one class of soils, and that the want of it produces the poverty of the other. Qualities that always thus accompany each other, cannot be otherwise than *cause* and *effect*. If further proof is wanting, it can be safely promised to be furnished when the practical application of calcareous manures to poor soils will be treated of, and their effects stated.

These deductions are then established as to all calcareous soils, and all poor soils — which descriptions comprise nine-tenths of all. This alone would open a wide field for the practical exercise of the truths we have reached. But still there remain strong objections and stubborn facts opposed to the complete proof of the proposition now under consideration, and consequently to the theory which that proposition is intended to support. The whole difficulty will be apparent at once when I now proceed to state that nearly all of our best soils, very little if at all inferior to the smaller portion of shelly lands, are as destitute of calcareous earth as the poorest. So far as I have examined, this deficiency is as general in the richest alluvial lands of the upper country — and what will be deemed by some as incredible, by far the greater part of the rich limestone soils between the Blue Ridge and Alleghany, are equally destitute of calcareous earth. These facts were not named before, to avoid embarrassing the discussion of other points — nor can they now be explained, and reconciled with my proposition, except through a circuitous and apparently digressive course of reasoning. They have not been kept out of view, nor slurred over, to weaken their force, and are now presented in all their strength. These difficulties will be considered, and removed, in the following chapters.

CHAPTER VI

CHEMICAL EXAMINATION OF RICH SOILS CONTAINING NO CALCAREOUS EARTH

PROPOSITION 2. *Continued.*

UNDER common circumstances, when any disputant admits facts that seem to contradict his own reasoning, such admission is deemed abundant evidence of their existence. But though placed exactly in this situation, the facts admitted by me are so opposed to all that scientific agriculturists have taught us to expect, that it is necessary for me to show the grounds on which my admission rests. Few would have believed in the absence of calcareous earth in all our poor soils — and far more strange is it that the same deficiency should extend to such rich soils as some that will be cited.

The following specimens, taken from well known and very fertile soils, were found to contain no calcareous earth. Many trials of other rich soils have yielded like results — and indeed, I have never found calcareous earth in any soil below the Falls, in which, or near which, some particles of shells were not visible.

1. Soil from Eppes' Island, which lies in James River near City Point; light and friable (but not silicious) brown loam, rich and durable. The surface is not many feet above the highest tides, and like most of the best river lands, this tract seems to have been formed by alluvion many ages ago, but which may be termed recent, when compared to the general formation of the tide-water district.

2. Black silicious loam from the celebrated lands on Back River, near Hampton.

3. Soil from rich land on Pocoson River, York county.

4. Black clay vegetable soil, from a fresh-water tide marsh on James River — formed by the most recent alluvion.

5. Alluvial soil of first rate fertility above the Falls of James River — dark brown clay loam, from the valuable and extensive body of bottom land belonging to General J. H. Cocke of Fluvanna.

The most remarkable facts of the absence of calcareous earth, are to be found in the limestone soils, between the Blue Ridge and Alleghany mountains. Of these, I will report all that I have examined, and none contained any calcareous earth, unless the contrary is stated.[1]

1 to 6. Limestone soils selected in the neighbourhood of Lexington, Virginia, by Professor Graham, with the view of enabling me to investigate this subject. All the specimens were from first rate soils, except one, which was from land of inferior value. One of the specimens, Mr. Graham's description stated to be "taken from a piece of land so rocky [with limestone] as to be unfit for cultivation — at least with the plough. I could scarcely select a specimen which I would expect to be more strongly impregnated with calcareous earth." — This specimen, by two separate trials, yielded only one grain of calcareous earth, from one thousand of soil. The other six soils contained none. The same result was obtained from

7. A specimen of alluvial land on North River, near Lexington.

8. Brown loam from the Sweet Spring valley, remarkable for its extraordinary productiveness and durability. It is of alluvial formation, and before it was drained, must have been

[1] Before the first of these trials was made, I supposed (as probably most other persons do,) that *limestone soil* was necessarily *calcareous*, and in a high degree. It is difficult to get rid of this impression entirely — and it may seem a contradiction in terms to say that a *limestone soil* is not calcareous. This I cannot avoid: I must take the term *limestone soil* as custom has already fixed it. But it should not be extended to any soils except those which are so near to limestone rock, as in some measure to be thereby affected in their qualities and value.

often covered and saturated by the Sweet Spring and other mineral waters, which hold lime in solution. The surrounding highland is limestone soil. Of this specimen, taken from about two hundred yards below the Sweet Spring, from land long cultivated every year, three hundred and sixty grains yielded not a particle of calcareous earth. It contained an unusually large proportion of *oxide of iron*, though my imperfect means enabled me to separate and collect only eight grains, the process evidently wasting several more.

About a mile lower down, drains were then making (in 1826) to reclaim more of this rich valley from the overflowing waters. Another specimen was taken from the bottom of a ditch just opened, eighteen inches below the surface. It was a black loam, and exhibited to the eye some very diminutive fresh-water shells, (perriwinkles, about one-tenth of an inch in length,) and many of their broken fragments. This gave, from two hundred grains, seventy-four of calcareous earth. But this cannot fairly be placed on the same footing with the other soils, as it had obviously been once the bottom of a stream, or lake, and the collection and deposit of so large and unusual a proportion of calcareous matter, seemed to be of animal formation. Both these specimens were selected at my request by one of our best farmers, and who also furnished a written description of the soils, and their situation.

9. High land in wood, west of Union, Monroe county. Soil, a black clay loam, lying on, but not intermixed at the surface with limestone rock. Subsoil, yellowish clay. The rock at this place, a foot below the surface. Principal growth, sugar maple, white walnut, and oak. This and the next specimen are from one of the richest tracts of highland that I have seen.

10. Soil similar to the last and about two hundred yards distant. Here the limestone showed above the surface, and the specimen was taken from between two large masses of fixed rock, and about a foot distant from each.

11. Black rich soil, from woodland between the Hot and

Warm Springs, in Bath county. The specimen was part of what was in contact with a mass of limestone.

12. Soil from the western foot of the Warm Spring Mountain, on a gentle slope between the Court House and the road, and about one hundred and fifty yards from the Warm Bath. Rich brown loam, containing many small pieces of limestone, but no finely divided calcareous earth.

13. A specimen taken two or three hundred yards from the last, and also at the foot of the mountain. Soil, a rich black loam, full of small fragments of limestone of different sizes, between that of a nutmeg and small shot. The land had never been broken up for cultivation. One thousand grains contained two hundred and forty grains of small stone or gravel, mostly limestone, separated mechanically, and sixty-nine grains of finely divided calcareous earth.

14. Black loamy clay, from the excellent wheat soil adjoining the town of Bedford in Pennsylvania: the specimen taken from beneath and in contact with limestone. One thousand grains yielded less than one grain of calcareous earth.

15. A specimen from within a few yards of the last, but not in contact with limestone, contained no calcareous earth: neither did the red clay subsoil, six inches below the surface.

16. Very similar soil, but much deeper, adjoining the principal street of Bedford — the specimen taken from eighteen inches below the surface, and adjoining a mass of limestone. A very small disengagement of gas indicated the presence of calcareous earth — but certainly less than one grain in one thousand, and perhaps not half that quantity.

17. Alluvial soil on the Juniata, adjoining Bedford —

18. Alluvial vegetable soil near the stream flowing from all the Saratoga mineral springs, and necessarily often covered and soaked by those waters, and

19. Soil taken from the bed of the same stream — neither contained any portion of carbonate of lime.

Thus it appears, that of nineteen specimens of soils, only

four contained calcareous earth, and three of these four, in exceedingly small proportions. It should be remarked that all these were selected from situations, which from their proximity to calcareous rock, or exposure to calcareous waters, were supposed most likely to present highly calcareous soils. If five hundred specimens had been taken without choice, from what are commonly limestone soils, (merely because they are not very distant from limestone rock, or springs of limestone water,) the analysis of that whole number would be less likely to show calcareous earth, than the foregoing short list. I therefore feel justified, from my own few examinations, and unsupported by any other authority, to pronounce that calcareous earth will very rarely be found in any soils between the falls of our rivers, and the navigable western waters. In a few specimens of some of the best soils from the borders of the Mississippi and its tributary rivers, I found calcareous earth present in all — but in small proportions, and in no case exceeding two per cent.

The foregoing details, respecting limestone lands, may perhaps be considered an unnecessary digression, in a treatise on the soils of the tide-water district. But the analysis of limestone soils furnishes the strongest evidence of the remarkable and novel fact of the general absence of calcareous earth — and the information thence derived, will be used to sustain the following steps of my argument. All the examinations of soils in this chapter concur in opposing the general application of the proposition that the deficiency of calcareous earth is the cause of the sterility of our soils. Having stated the objection in all its force, I shall now proceed to inquire into its causes, and endeavour to dispel its apparent opposition to my doctrine.

CHAPTER VII

PROOFS OF THE EXISTENCE OF ACID
AND NEUTRAL SOILS

Proposition 2. *Continued.*

Sufficient evidence has been adduced to prove that many of our most fertile and valuable soils are destitute of calcareous earth: but it does not necessarily follow that such has always been their composition — or that they may not now contain lime combined with some other acid than the carbonic. That this is really the case, I shall now offer proofs to establish — and not only maintain this position with regard to those valuable soils, but shall contend that lime in some proportion, combined with *vegetable acid*, is present in every soil capable of supporting vegetation.

But while I shall endeavour to maintain these positions, without asking or admitting any exception, let me not be understood as asserting that the original ingredient of calcareous earth was always the sole cause of the fertility of any particular soil, or that a knowledge of the proportion contained, would serve to measure the capacity of the soil for improvement. Calcareous soils not differing materially in qualities or value, often exhibit a remarkable difference in their respective proportions of calcareous earth: so that it would seem, that a small quantity, aided by some other unknown agent, may give as much capacity for improvement, and ultimately produce as much fertility, as ten times that proportion, under other circumstances.

In all naturally poor soils, producing freely, in their virgin state, pine and whortleberry, and sorrel after cultivation, I suppose to have been formed some *vegetable acid*, which, after taking up whatever small quantity of lime might have been present, still remains in excess in the soil, and nourishes in the highest degree the plants named above, but is a poison to all useful crops; and effectually prevents such soils becoming rich, from either natural or artificial applications of putrescent manures.

In a *neutral soil*, I suppose calcareous earth to have been sufficiently abundant to produce a high degree of fertility — but that it has been decomposed, and the lime taken up, by the gradual formation of vegetable acid, until the lime and the acid neutralize and balance each other, leaving no considerable excess of either. Such are all our fertile soils that are not calcareous.

These suppositions remain to be proved, in all their parts.

No opinion has been yet advanced that is less supported by good authority, or to which more general opposition may be expected, than that which supposes the existence of acid soils. The term *sour soil* is frequently used by farmers, but in so loose a manner as to deserve no consideration: it has been thus applied to any cold and ungrateful land, without intending that the term should be literally understood, and perhaps without attaching to its use any precise meaning whatever. Dundonald only, of all those who have applied chemistry to agriculture, has asserted the existence of vegetable acid in soils: [1] but he has offered no analysis, nor any other evidence to establish the fact — and his opinion has received no confirmation, nor even the slightest notice, from later and more able investigators of the chemical characters of soils. Kirwan and Davy profess to enumerate all the common ingredients of soils, and it is not intimated by either, that vegetable acid is one of them. Even this tacit denial by Davy, more strongly opposes the existence of vegetable acid, than it is supported

---

[1] Dundonald's Connexion of Chemistry and Agriculture. [Archibald Cochrane (Earl of Dundonald), *A Treatise Showing the Intimate Connection . . . between Agriculture and Chemistry* (London, 1795).]

by the opinion of Dundonald, or any of those writers on agriculture who have admitted its existence. Grisenthwaite,[2] a
late writer on agricultural chemistry, and who has the advantage of knowing the discoveries, and comparing the opinions
of all his predecessors, expressly denies the possibility of any
acid existing in soils. His "New Theory of Agriculture" contains the following passage: "Chalk has been recommended as a
substance calculated to correct the sourness of land. It would
surely have been a wise practice to have previously ascertained
this existence of acid, and to have determined its nature, in
order that it might be effectually removed. The fact really is,
that no soil was ever yet found to contain any notable quantity of acid. The acetic and the carbonic are the only two that
are likely to be generated by any spontaneous decomposition
of animal or vegetable bodies, and neither of them have any
fixity when exposed to the air." Thus, then, my doctrine is
deprived of even the feeble support it might have had from
Dundonald's mere opinion, if that opinion had not been contradicted by later and better authority: and the only support
that I can look for, will be in the facts and arguments that I
shall be able to adduce.

I am not prepared to question what Grisenthwaite states
as a chemical fact, "that no soil was ever yet found to contain
any notable quantity of acid." No soil examined by me for
this purpose, gave any evidence of the presence of uncom-
bined acid. Still, however, the term acid may be applied with
propriety to soils, in which growing vegetables continually
receive acid from the decomposition of others, (for which no
"fixity" is requisite,) or in which acid is present, not free, but
combined with some base, by which it is readily yielded to
promote, or retard, the growth of plants in contact with it.
It will be sufficient for my purpose to show that certain soils
contain some substance, or possess some quality, which pro-

[2] [William Grisenthwaite, *A New Theory of Agriculture* . . . (Wells,
1819).]

motes almost exclusively the growth of acid plants — that this power is strengthened by adding known vegetable acids to the soil — and is totally removed by the application of calcareous manures, which would necessarily destroy any acid, if it were present. Leaving it to chemists to determine the nature and properties of this substance, I merely contend for its existence and effects: and the cause of these effects, whatever it may be, for the want of a better name, I shall call *acidity*.

The proofs now to be offered in support of the existence of acid and neutral soils, however weak each may be when considered alone, yet when taken in connexion, will together form a body of evidence not easily to be resisted.

1st Proof. Pine and common sorrel have leaves well known to be acid to the taste: and their growth is favoured by the soils which I suppose to be acid, to an extent which would be thought remarkable in other plants on the richest soils. Except wild locust on the best river land, no growth can compare in rapidity with pines on soils naturally poor, and even greatly reduced by long cultivation. Pines usually stand so thick on old exhausted fields, that the increase of size in each plant is greatly retarded — but if the whole growth of an acre is estimated, it would probably exceed in quantity that of the richest soils, of the same age and on the same space. Every cultivator of corn on poor light soil knows how rapidly sorrel [3] will cover his otherwise naked field, unless kept in check by continual tillage — and that to root it out, so as to prevent the like future labour, cannot be effected by any mode of cultivation whatever. This weed too is considered far more hurtful to growing crops, than any other of equal size. Yet neither of these acid plants can thrive on the best lands. Sorrel cannot

---

[3] *Rumex acetosa.* The wood sorrel (*oxalis acetocella*) is of a very different character. This prefers rich and calcareous soils, and I have seen it growing on places calcareous to excess. It would seem, therefore, that wood sorrel forms its acid from the atmosphere, and does not draw it from the soil, as is evidently the case with common sorrel.

even live on a calcareous soil — and if a pine is sometimes found there, it has nothing of its usual elegant form, but seems as stunted and ill shaped as if it had always suffered for want of nourishment. Innumerable facts, of which these are examples, prove that these acid plants must derive from their favourite soil some kind of food peculiarly suited to their growth, and quite useless, if not hurtful, to cultivated crops.

2. Dead acid plants are the most effectual in promoting the growth of living ones. When pine leaves are applied to a soil, whatever acid they contain is of course given to that soil, for such time as circumstances permit it to retain its form, or peculiar properties. Such an application is often made on a large scale, by cutting down the second growth of pines, on land once under tillage, and suffering them to lie a year before clearing and cultivating the land. The invariable consequence of this course, is a growth of sorrel for one or two crops, so abundant and so injurious to the crops, as to more than balance any benefit derived by the soil, from the vegetable matter having been allowed to rot. From the general experience of this effect, most persons tend pine land as soon as cut down, after carefully burning the whole of the heavy cover of leaves, both green and dry. Until within a few years, it was generally supposed that the leaves of pine were worthless, if not hurtful, in all applications to cultivated land — which opinion doubtless was founded on such facts as have been just stated. But if they are used as litter for cattle, and heaped to ferment, the injurious quality of pine leaves is destroyed, and they become a valuable manure. This practice is but of recent origin — but is highly approved, and rapidly extending.

On one of the washed and barren declivities (or *galls*) which are so numerous on all our farms, I had the small gullies packed full of green pine bushes, and then covered with the earth drawn from the equally barren intervening ridges, so as nearly to smooth the whole surface. The whole piece bore nothing previously except a few scattered tufts of poverty grass, and dwarfish sorrel, all of which did not prevent the spot seeming quite bare at midsummer, if viewed at the distance of one hundred yards. This operation was performed in February or March. The land was not cultivated, nor again observed until the second summer afterwards. At that time, the piece remained as bare as formerly, except along the filled gullies, which throughout the whole of their crooked courses, were covered by a thick and tall growth of sorrel, remarkably luxuriant for any situation, and which being bounded exactly by the width of the narrow gullies, had the appearance of some vegetable sown thickly in drills, and kept clean by tillage. So great an effect of this kind has not been produced within my knowledge — though facts of like nature and leading to the same conclusion, are of frequent occurrence. If small pines standing thinly over a broom grass old field, are cut down and left to lie, under every top will be found a patch of sorrel, before the leaves have all rotted.

3. The growth of sorrel is not only peculiarly favoured by the application of vegetables containing acids already formed, but also by such matters as will form acid in the course of their decomposition. Farm-yard manure, and all other putrescent animal and vegetable substances, form *acetic acid* as their decomposition proceeds.[4] If heaps of rotting manure are left without being spread, in a field the least subject to produce sorrel, a few weeks of growing weather will bring out that plant close around every heap — and for some time, it will continue to show more benefit from that rank manuring than any other grass. For several years my winter-made manure was spread and ploughed in on land not cultivated until the next autumn, or the spring after. This practice was founded on the mistaken opinion, that it would prevent much of the usual exposure to evaporation and waste of the manure. One of the reasons which alone would have compelled me to abandon this absurd practice, was, that a crop of sorrel always

---

[4] *Agr. Chem.*, p. 187.

followed, (even on good soils that before barely permitted a scanty growth to live,) which so injured the next grain crop as greatly to lessen the benefit from the manure. Sorrel unnaturally produced by such applications, does not infest the land longer than until we may suppose the acid to be removed by cultivation and other causes.

It may be objected that my authorities prove only the formation of a single vegetable acid in soil, the acetic — that my facts show only the production of a single acid plant, sorrel — and that the acid which sorrel contains is not the acetic, but the oxalic.[5] From the application of acids to recently ploughed land, no acid plant except sorrel is made to grow, because that only can spring up speedily enough to arrest the fleeting nutriment. Poverty grass grows only on the same kinds of soil, and generally covers them after they have been a year free from a crop, but does not show sooner — and broom grass and pines require two years before their seeds will produce plants. But when pines begin to spread over the land, they soon put an end to the growth of all other plants, and are abundantly supplied with their acid food, from the dropping of their own leaves. Thus they may be first supplied with the vegetable acid ready formed in the leaves, and afterwards with the acetic acid, formed by their subsequent slow decomposition. It does not weaken my argument, that the product of a plant is a vegetable acid different from the one supposed to have nourished its growth. All vegetable acids (except the prussic) however different in their properties, are composed of the same three elementary bodies, differing only in their proportions [6] — and consequently are all resolvable into each other. A little more, or a little less of one or the other of these ingredients, may change the acetic to the oxalic acid, and that to any other. We cannot doubt but that such simple changes may be produced by the chemical

[5] *Ibid.*, Lecture 3.
[6] Carbon, Oxygen and Hydrogen, *ibid.*, Lecture 3, p. 78.

powers of vegetation, when others are effected, far more difficult for us to comprehend. The most tender and feeble organs, and the mildest juices, aided by the power of animal or vegetable life, are able to produce decompositions and combinations, which the chemist cannot explain, and which he would in vain attempt to imitate.

4. This ingredient of soils which nourishes acid plants, also poisons cultivated crops. Plants have not the power of rejecting noxious fluids, but take up by their roots every thing presented in a soluble form.[7] Thus the acid also enters the sap-vessels of cultivated plants, stunts their growth, and makes it impossible for them to attain that size and perfection, which their proper food would ensure, if it was presented to them without its poisonous accompaniment. When the poorest virgin woodland is cut down, it is covered and filled to excess with leaves and other rotted and rotting vegetable matters. Can a heavier vegetable manuring be desired? And as it completely rots during cultivation, must not it offer to the growing as abundant a supply of food as they can require? — Yet the best product obtained may be from ten to fifteen bushels of corn, or five or six of wheat, soon to come down to half those quantities. If the noxious quality which causes such injury is an acid, it is as certain as any chemical truth whatever, that it will be neutralized, and its powers destroyed, by applying enough of calcareous earth to the soil: and precisely such effects are found whenever that remedy is tried. On land thus relieved of this unceasing annoyance, the young corn no longer appears of a pale and sickly green, approaching to yellow, but takes immediately a deep healthy colour, by which it may readily be distinguished from any remaining in its former state, before there is any perceptible difference in size. The crop will produce fifty to one hundred per cent. more, the first year, before its supply of food can possibly have been increased — and the soil is soon found not only

[7] Ibid., Lecture 6, p. 186.

cleared of sorrel, but incapable of producing it. I have anticipated these effects of calcareous manures, but they will hereafter be established beyond contradiction.

5. The truth of the existence of either acid, or neutral
soils, depends on the existence of the other — and to prove
either, will necessarily establish both. If acid exists in soils,
then wherever it meets with calcareous earth, the two substances must combine and neutralize each other, so far as their
proportions are properly adjusted. On the other hand, if I
can show that compounds of lime and vegetable acid are
present in most soils, it follows inevitably that nature has
provided means by which soils can generally obtain this
acid: and if the amount formed can balance the lime, the
operation of the same causes can exceed that quantity, and
leave an excess of free acid. From these premises will be
deduced the following proofs.

It has been stated (page 39) that the process recommended
by chemists for finding the calcareous earth in soils was unfit
for that purpose, because a *precipitate* was always obtained
even when no calcareous earth, or carbonate of lime was
present. Frequent trials have shown me that this precipitate
is considerably more abundant from good soils than bad. The
substance thus obtained from rich soils by solution and precipitation, in every case that I have tried, contains some calcareous earth, although the soil from which it was derived
had none. The alkaline liquor from which the precipitate has
been separated, we are told will, after boiling, let fall the
carbonate of magnesia, if any had been in the soil: but when
any notable deposit is thus obtained, it will often be found
to consist more of carbonate of lime, than of magnesia. The
following are examples of such products:

One thousand grains of tide marsh soil (described page 47)
acted on by muriatic acid in the pneumatic apparatus, gave
out no carbonic acid gas, and therefore could have contained
no carbonate of lime. The precipitate obtained from the same

weighed sixteen grains — which being again acted on by
sulphuric acid, evolved as much gas as showed that three
grains had been converted to carbonate of lime.

Two hundred grains of alluvial soil from Saratoga Springs
(page 49, No. 18,) containing no carbonate of lime, yielded
a precipitate of twelve grains, of which three was carbonate
of lime — and a deposit from the alkaline solution weighing
six grains, four of which was carbonate of lime.

Seven hundred grains of limestone soil from Bedford (part
of the specimen marked 14, page 49,) contained about two-
thirds of a grain of carbonate of lime — and its precipitate of
twenty-eight grains, only yielded two grains: but the alkaline
solution deposited eleven grains of the carbonates of lime and
magnesia, of which at least five was of the former, as there
remained seven and a half of solid matter, after the action of
sulphuric acid.[8]

From this process, there can be no doubt but that the soil
contained a proportion of some *salt of lime* (or lime combined with some kind of acid) which being decomposed by
and combined with the muriatic acid, was then precipitated,
not in its first form, but in that of carbonate of lime — it being
supplied with carbonic acid from the carbonate of potash,
used to produce the precipitation. The proportions obtained
in these cases were small; but it does not follow that the whole
quantity of lime contained in the soil was found. However,
to the extent of this small proportion of lime, is proved clearly

[8] The measurement of the carbonic acid gas evolved, was relied on to
show the whole amount of *carbonates* present — and sulphuric acid was
used to distinguish between lime and magnesia, in the deposit from the
alkaline solution. If any alumine or magnesia had made part of the solid
matter exposed to diluted sulphuric acid, the combinations formed would
have been soluble salts, which would of course have remained dissolved and
invisible in the fluid. Lime only of the four earths forms with sulphuric acid
a substance but slightly soluble, and which therefore can be mostly separated
in a solid form. The whole of this substance (sulphate of lime) cannot be
obtained in this manner, as a part is always dissolved: but whatever is
obtained, proves that at least two-thirds of that quantity of carbonate of
lime had been present.

the presence of enough of some acid (and that not the carbonic) to combine with it. Neither could it have been the sulphuric, or the phosphoric acid: for though both the sulphate and phosphate of lime are in some soils, yet neither of these salts can be decomposed by muriatic acid.

6. The strongest objection to the doctrine of neutral soils is, that if true, the salt formed by the combination of the lime and acid must often be present in such large proportions, that it is scarcely credible that its presence and nature should not have been discovered by any of the chemists who have analyzed soils. This difficulty I cannot remove: but it may be met (or neutralized, to borrow a figure from my subject,) by showing that an equal difficulty awaits those who may support the other side of the argument. The theory of geologists of the formation of soils from the decomposition, or disintegration of rocks, is received as true by scientific agriculturists. The soils thus supposed to be formed, receive admixtures from each other, by means of different operations of nature, and after being more or less enriched by the decay of their own vegetable products, make the endless variety of existing soils.[9] But where a soil lying on, and thus formed from any particular kind of rock, is so situated that it could not have been moved, or received considerable accessions from torrents, or other causes, then, according to this theory, the rock and the soil should be composed of the same materials — and such soils as the specimens marked 11 and 16 (pages 48 and 49) would be, like the rock they touched, nearly pure calcareous earth. Such are the doctrines received and taught by Davy, or the unavoidable deductions from them. But without contending for the full extent of this theory of the formation of soils, (because I consider it almost entirely false,) every one must admit that soils thus situated, must have received in the lapse of ages, some accessions to

[9] *Agr. Chem.*, p. 131. Also "Treatise on Agriculture" (by General Armstrong) in vol. i [1819] of *American Farmer* [pp. 74f].

their bulk, from the effects of frost, rain, sun, and air, on the limestone in contact with them. All limestone soils, properly so called, exhibit certain marked and peculiar characters of colour, texture, and products, which can only be derived from receiving into their composition more or less of the rock which lies beneath, or rises above their surface. This mixture will not be denied by any one who has observed limestone soils, and reasons fairly, whether his investigation begins with the causes or their effects. If then all this gain of calcareous earth remains in the soil, why is none, or almost none, discovered by accurate chemical analysis? Or, if it be supposed not present, nor yet changed in its chemical character, in what possible manner could a ponderous and insoluble earth have made its escape from the soil? To remove this obstacle without admitting the operation of acid in making such soils neutral, will be attended with at least as much difficulty, as any arising from that admission being made.

7. But we are not left entirely to conjecture that soils were once more calcareous than they now are, if chemical tests can be relied on to furnish proof. Acid soils that have received large quantities of calcareous earth as manure, after some time, will yield very little when analyzed. To a soil of this kind, full of vegetable matter, I applied in 1818 and 1821 fossil shells at such a known and heavy rate as would have given to the soil (by calculation) at least three per cent. of calcareous earth, for the depth of five inches. Only a small portion of the shelly matter was finely divided when applied. Since the application of the greater part of this dressing, (only one-fourth having been laid on in 1818,) no more than six years had passed before the following examinations were made — and the cultivation of five crops in that time, three of which were horse-hoed, must have well mixed the calcareous earth with the soil. Three careful examinations gave the following results.

No. 1. — 1000 grains yielded 7½ of coarse calcareous earth,
And less than   ½ of finely divided.

                      ———
                       8
                      ———
No. 2. — 1000 grains   5 coarse,
                            2 finely divided.

                      ———
                       7
                      ———
No. 3. — 1500 grains   15 coarse,
                            2½ finely divided.

                      ———
                      17½
                      ———

The specimens No. 1 and No. 2 were obtained by handfuls
of soil from several places, (four in one case, and twelve in
the other) mixing them well together, and then taking the
samples for trial from the two parcels. On such land, when
not recently ploughed, there will always be an over proportion of the pieces of shells on the surface, as the rains have
settled the fine soil, and left exposed the coarse matters. On
this account, in making these two selections, the upper half
inch was first thrown aside, and the handful dug from below.
No. 3 was taken from a spot showing a full average thickness
of shells, and included the surface. I considered the three
trials made as fairly as possible, to give a general average.
Small as is the proportion of finely divided calcareous earth
exhibited, it must have been increased by rubbing some particles from the coarse fragments, in the operation of separating
them by a fine sieve. Indeed it may be doubted whether any
proportion remained very finely divided — or in other words,
whether it was not combined with acid, as fast as it was so
reduced. But without the benefit of this supposition, the finely
divided calcareous earth in the three specimens, averaged only
one and one-fourth grains to the thousand, which is one
twenty-fourth of the quantity laid on: and the total quantity
obtained, of coarse and fine, is eight grains in one thousand, or

about one-fourth of the original proportion. All the balance had changed its form, or otherwise disappeared in the few years that had passed since the application.

The very small proportions of finely divided calcareous earth compared to the coarse, in some shelly soils, furnish still stronger evidence of this kind. Of the York River soil, (described page 43, No. 5,)
1260 grains, yielded of coarse calcareous

parts,	-	-	-	-	-	-	168 grains.
And of finely divided,	-	-	-	-	8

1044 of the rich Nansemond soil, (No. 6,)   544 coarse.
                                                  18 fine.

As many of the shells and their fragments in these soils are in a mouldering state, it is incredible that the whole quantity of finely divided particles derived from them should have amounted to no more than these small proportions. Independent of the action of natural causes, the plough alone, in a few years, must have pulverized at least as much of the shells.

8. In other cases where the operations of nature have been applying calcareous earth, for ages, none now remains in the soil; and the proof thence derived is more striking, than any obtained from artificial applications, of a few years standing. Valleys subject to be frequently overflowed and saturated by the water of limestone streams, must necessarily retain a new supply of calcareous earth from every such soaking and drying.

Limestone water contains the *super-carbonate of lime*, which is soluble: but this loses its excess of carbonic acid when left dry by evaporation, and becomes the carbonate of lime, which not being soluble, is in no danger of being removed by subsequent floods. Thus accessions are slowly but continually made, through many centuries. Yet such soils are found containing no calcareous earth — of which a remarkable example is presented in the soil of the cultivated part of the Sweet Spring Valley, (No. 8, page 47.) [10]

9. All *wood ashes* contain salts of lime, (and most kinds in large proportions,) which could have been derived from no other source than the soil on which the trees grew. The lime thus obtained is principally combined with carbonic acid, and partly with the phosphoric, forming phosphate of lime. The table of Saussure's analyses of the ashes of numerous plants,[11] is sufficient to show that these products are general, if not universal. The following examples of some of my own examinations, prove that ashes yield calcareous earth in proportions suitable to their kind, although the wood grew on soils destitute of that ingredient — as was ascertained with regard to each of these soils.

100 *grains of ashes from*	*What soil taken from*	Carbonate of Lime.	Phosphate of Lime.
Whortleberry bushes, the entire plants, except the leaves,	Acid silicious loam,	4 grains.	4 grains.
Equal parts of the bark, heart, and sapwood, of an old locust,	The same,	51	18
Young locust bushes entire,	Rich neutral clay loam,	40	30
Young pine bushes,	Acid silicious loam,	9	6
Body of a young pine tree,	Acid clay soil,	14	18

The potash was first carefully taken out of all these samples. The remaining solid matter was silicious sand, and charcoal:

[10] The excess of carbonic acid which unites with lime and renders the compound soluble in water, is lost by exposure of the calcareous water to the air, as well as by evaporation to dryness. [Frederick Accum, *System of Theoretical and Practical Chemistry*, 2 vols., 2nd ed. (Philadelphia, 1814), II, 241–248.] The masses of soft calcareous rock which are deposited in the rapids of limestone streams, are examples of the loss of carbonic acid from exposure to the air; and the stalactites in caves, the deposit of the slow-dropping calcareous water, are examples of the same effect produced by evaporation. A similar deposit of insoluble carbonate of lime, from both these causes, is necessarily made on all land subject to be overflowed by limestone waters.

[11] Quoted in *Agr. Chem.*, Lecture 3.

the proportion of the latter varying according to the degree
of heat used in burning the wood, which was not permitted
to be very strong, for fear of converting the calcareous earth
into *quick-lime*.

All the *carbonate of lime* yielded by ashes, was necessarily
furnished in some form by the soil on which the plants grew
— and when the soil itself contained no *carbonate*, some other
compound of lime must have been present, to enable us to
account for the certain and invariable results. The presence
of a combination of lime with some *vegetable* acid, and none
other, would serve to produce such effects. According to
established chemical laws, if any such combination had been
taken up into the sap-vessels of the tree, it would be decomposed by the heat necessary to convert the wood to ashes;
the acid would be reduced to its elementary principles, and
the lime would immediately unite with the carbonic acid,
(which is produced abundantly by the process of combus-
tion,) and thus present a product of *carbonate of lime* newly
formed from the materials of the other substances decomposed.[12]

On the foregoing facts and deductions, I am content to
rest the truth of the existence of acid and neutral soils. Supposing the doctrine to be sufficiently proved, it may be useful
to trace the formation of acidity in different soils, according
to the views which have been presented, and to display the
promise which that quality holds out for improving those
soils, which it has hitherto rendered barren and worthless.

Every neutral soil at some former time contained calcareous
earth in sufficient quantity to produce the uniform effect of
that ingredient, of storing up and fixing fertility. The decomposition of the successive growths of plants left to rot on
the rich soil, continually formed vegetable acid, which slowly

[12] The reasoning on the presence of the carbonate of lime found in ashes
from acid soils, does not apply to the phosphate of lime which is also always
present. The latter salt is not decomposed by any known degree of heat,
(Art. "Chemistry," in *Edin. Ency.*) and therefore might have remained unchanged, in passing from the soil to the tree, and thence to the ashes.

and gradually united with the lime in the soil. At last these two principles balanced each other, and the soil was no longer calcareous, but became neutral. Instead of its former ingredient carbonate of lime, it was now supplied with a *vegetable salt of lime*. This change of soil does not affect the natural growth, which remains the same, and thrives as well as when the soil was calcareous — and when brought into cultivation, the soil is equally productive under all crops suited to calcareous soils. If the supplies of vegetable matter continue, the soil may even become acid in some measure, as may be evidenced by the growth of sorrel — but without losing any of its fertility before acquired. The quantity of acidity in any soil frequently varies: it is increased by the growth of such plants as delight to feed on it, and by the decomposition of all vegetable matters. Hence the longer a poor field remains at rest, and not grazed, the more acid it becomes — and this evil keeping pace with the benefits derived, is the cause why so little improvement, or increased product, is obtained from putting acid soils under that mild treatment. Cultivation not only prevents new supplies, but also diminishes the acidity already present in excess, by exposing it to the atmosphere — and the more a soil is exhausted, the more will its acidity be lessened.

We have seen that even acid soils contain a little salt of lime, and therefore must have once been slightly calcareous. Indeed it may be well doubted whether any soil destitute of lime in any form, would not soon become a perfect barren, incapable of producing a spire of grass. But such small proportions of calcareous earth were soon equalled, and then exceeded, by the formation of vegetable acid, before much productiveness was caused. The soil being thus changed, the plants suitable to calcareous soils died off, and gave place to others which produce, as well as feed and thrive on acidity. Still, however, even these plants furnish abundant supplies of vegetable matter, sufficient to enrich the land in the highest degree: but the antiseptic power of the acid prevents the leaves from rotting

for years, and even then the soil has no power to profit by
them. Though continually wasted, the vegetable matter is
always present in abundance; but must remain almost useless
to the soil, until the accompanying acidity shall be destroyed.

Nearly all the woodland now remaining in Lower Virginia,
and much of what has long been arable, is rendered unproductive by acidity, and successive generations have toiled on
them without remuneration, and without suspecting that their
worst virgin land was then richer than their manured lots
appeared to be. The cultivator of such soil, who knows not
its peculiar disease, has no other prospect than a gradual decrease of his always scanty crops. But if the evil is once understood, and the means of its removal within his reach, he has
reason to rejoice that his soil was so constituted as to be preserved from the effects of the improvidence of his forefathers,
who would have worn out any land not almost indestructible.
The presence of acid, by restraining the productive powers
of the soil, has in a great measure saved it from exhaustion;
and after a course of cropping which would have utterly
ruined soils much better constituted, the powers of our acid
land remain not greatly impaired, though dormant, and ready
to be called into action by merely being relieved of its acid
quality. A few crops will reduce a new acid field to such a
low product that it scarcely will pay for its cultivation — but
no great change is afterwards caused, by continuing scourging
tillage and grazing, for fifty years longer. Thus our acid soils
have two remarkable and opposite qualities, both proceeding
from the same cause: they cannot be enriched by manure, nor
impoverished by cultivation, to any great extent. Qualities
so remarkable deserve all our powers of investigation: yet
their very frequency seems to have caused them to be over-
looked — and our writers on agriculture have contrived to
urge those who seek improvement to apply precepts drawn
from English authors, to soils which are totally different from
all those for which their instructions were intended.

CHAPTER VIII

THE MODE OF OPERATION OF CALCAREOUS EARTH IN SOILS

PROPOSITION 3. *The fertilizing effects of calcareous earth are chiefly produced by its power of neutralizing acids, and of combining putrescent manures with soils, between which there would otherwise be but little if any chemical attraction.*

PROPOSITION 4. *Poor and acid soils cannot be improved durably, or profitably, by putrescent manures, without previously making them calcareous, and thereby correcting the defect in their constitution.*

IT has already been made evident that the presence of calcareous earth in a natural soil causes great and durable fertility: but it still remains to be determined, to what properties of this earth its peculiar fertilizing effects are to be attributed.

Chemistry has taught that silicious earth, in any state of division, attracts but slightly, if at all, any of the parts of putrescent animal and vegetable matters.[1] But even if any slight attraction really exists when the earth is minutely divided, for experiments in the laboratory of the chemist, it cannot be exerted by silicious sand in the usual form in which nature gives it to soils — that is, in particles comparatively coarse, loose, and open, and yet each particle impenetrable to any liquid, or gaseous fluid that might be passing through the vacancies. Hence, silicious earth can have no power, chemical or mechanical, either to attract enriching manures, or to preserve them when actually placed in contact: and soils in which the qualities of this earth greatly predominate, must give out freely all they have received, not only to a growing crop,

[1] *Agr. Chem.*, p. 129.

but to the sun, air, and water, so as soon to lose the whole. No portion of putrescent matter can remain longer than the completion of its decomposition — and if not arrested during this process, by the roots of living plants, all will escape in the form of *gas*, into the air, without leaving a trace of lasting improvement. With a knowledge of these properties, we need not resort to the common opinion that manure sinks through sandy soils, to account for its rapid disappearance.[2]

Aluminous earth, by its closeness, mechanically excludes those agents of decomposition, heat, air, and moisture, which sand so freely admits; and therefore soils in which this earth predominates give out manure much more slowly than sand, whether for waste or for use. The practical effect of this is universally understood — that clay soils retain manure much longer than sand, but require much heavier applications to

[2] Except the very small proportions of earthy, saline and metallic matters that may be in animal and vegetable manures, the whole balance of their bulk (and the whole of whatever can feed plants,) is composed of different elements, which are known only in the form of *gasses* — into which they must be finally resolved, after going through all the various stages of fermentation and decomposition. So far from sinking in the earth, these final results could not be possibly confined there, but must escape into the atmosphere as soon as they take a gaseous form, unless immediately taken up by the organs of growing plants. It is probable that but a small portion of any dressing of manure remains long enough in the soil to make this final change — and that nearly all is used by growing plants, during previous changes, or carried off by air and water. During the progress of the many changes caused by fermentation and decomposition, every soluble product may certainly sink as low as the rains penetrate: but it cannot descend lower than the water, and that, together with the soluble manure, will be again drawn up by the roots of plants. Should the soil need draining, to take off water passing beneath the surface, the soluble manure might perhaps be carried off by those springs. We as yet are but little informed as to the particular changes made, and the various new substances successively formed and then decomposed, during the whole duration of putrescent manures in the soil — and no field for discovery would better reward the investigations of the agricultural chemist. For want of this knowledge we proceed at random in using manures, instead of being enabled to conform to any rule founded on scientific principles: nor can we hope so to manage manures with regard to their fermentation, the time and manner of application, mixing with other substances, &c., as to enable the crops to seize every enriching result as it is produced, and to postpone as long as possible the final results of decomposition.

show as much effect at once. But as this means of retaining manure is altogether mechanical, it serves only to delay both its use and its waste. Aluminous earth also exerts some chemical power in attracting and combining with manures, but too weakly to enable a clay soil to become rich by natural means.

Davy states that both aluminous and calcareous earth will combine with any vegetable extract, so as to render it less soluble, and consequently not subject to the waste that would otherwise take place, and hence that "the soils which contain most alumina and carbonate of lime, are those which act with the greatest chemical energy in preserving manures." Here is high authority for calcareous earth possessing the power which my subject requires, but not in so great a degree as I think it deserves. Davy apparently places both earths in this respect on the same footing, and allows to aluminous soils retentive powers equal to the calcareous. But though he gives evidence (from chemical experiments) of this power in both earths, he does not seem to have investigated the difference of their forces. Nor could he deem it very important, holding the opinion which he elsewhere expresses, that calcareous earth acts "merely by forming a useful earthy ingredient in the soil," and consequently attributing to it no remarkable chemical effects as a manure. I shall offer some reasons for believing that the powers of attracting and retaining manure, possessed by these two earths, differ greatly in force.

Our aluminous and calcareous soils, through the whole of their virgin state, have had equal means of receiving vegetable matter; and if their powers for retaining it were nearly equal, so would be their acquired fertility. Instead of this, while the calcareous soils have been raised to the highest condition, many of the tracts of clay soil remain the poorest and most worthless. It is true that the one laboured under acidity, from which the other was free. But if we suppose nine-tenths of the vegetable matter to have been rendered useless by that poisonous quality, the remaining tenth, applied for so long a time, would have made any soil fertile, that had the power to retain the enriching matter.

Calcareous earth has power to preserve those animal matters which are most liable to waste, and which give to the sense of smell full evidence when they are escaping. Of this, a striking example is furnished by an experiment which was made with care and attention. The carcass of a cow that was killed by accident in May, was laid on the surface of the earth, and covered with about seventy bushels of finely divided fossil shells and earth, (mostly silicious,) their proportions being as thirty-six to sixty-four. After the rains had settled the heap, it was only six inches thick over the highest point of the carcass. The process of putrefaction was so slow, that several weeks passed before it was over; nor was it ever so violent as to throw off any effluvia that the calcareous earth did not intercept in its escape, so that no offensive smell was ever perceived. In October, the whole heap was carried out and applied to one-sixth of an acre of wheat — and the effect produced far exceeded that of the calcareous manure alone, which was applied at the same rate on the surrounding land. No such power as this experiment indicated, will be expected from clay.

Quicklime is used to prevent the escape of offensive effluvia from animal matter; but its operation is entirely different from that of calcareous earth. The former effects its object by "eating" or decomposing the animal substance, (and nearly destroying it as manure,) before putrefaction begins. The operation of calcareous earth is to moderate and retard, but not to prevent putrefaction — not to destroy the animal matter, but to preserve it effectually by forming new combinations with the products of putrefaction. This mode of using calcareous earth might be practiced to great advantage near towns, where carcasses and other animal matters are so abundant as to be a public nuisance. [Appendix. E.]

The power of calcareous earth to combine with and retain putrescent manures, implies the power of fixing them in any soil to which both are applied. The same power will be equally exerted if the putrescent manure is applied to a soil which had previously been made calcareous, whether by nature, or by art. When a chemical combination is formed between the two kinds of manure, the one is necessarily as much fixed in the soil as the other. Neither air, sun, or rain, can then waste the putrescent manure, because neither can take it from the calcareous earth, with which it is chemically combined. Nothing can effect the separation of the parts of this compound manure, except the attractive power of growing plants — which as all experience shows, will draw their food from this combination as fast as they require it, and as easily as from sand. The means then by which calcareous earth acts as an improving manure, are, *completely preserving putrescent manures from waste, and yielding them freely for use.* These particular benefits, however great they may be, cannot be seen very quickly after a soil is made calcareous, but will increase with time, and the means for obtaining vegetable matters, until their accumulation is equal to the soil's power of retention. The kind, or the source of enriching manure, does not alter the process described. The natural growth of the soil, left to die and rot, or other putrescent manures collected and applied, would alike be seized by the calcareous earth, and fixed in the soil.

This, the most important and valuable operation of calcareous earth, gives nothing to the soil — but only secures the other manures, and gives *them* wholly to the soil. In this respect, the action of calcareous earth on soils, is precisely like that of *mordants* in "setting" or fixing colours. When alum, for example, is used by the dyer for this purpose, it adds not the slightest tinge of itself — but it holds to the cloth, and also to the otherwise fleeting dye, and thus fixes them permanently together. Without the mordant, the colour might have been equally vivid, but would be lost by the first wetting of the cloth. [Appendix. F.]

The next most valuable property of calcareous manures for the improvement of soil, is their *power of neutralizing acids*, which has already been incidentally brought forward in the preceding chapter. According to the views already presented, our poorest cultivated soils contain more vegetable matter than they can beneficially use — and when first cleared, have it in great excess. So antiseptic is the acid quality of poor woodland, that before the crop of leaves of one year can entirely rot, two or three others will have fallen — and there are always enough, at any one time, to greatly enrich the soil, if the leaves could be rotted and fixed in it, at once.[3] The presence of acid, by preventing or retarding putrefaction, keeps the vegetable matter inert, and even hurtful on cultivated land; and the crops are still further injured, by taking up the poisonous acid, with their nutriment. A sufficient quantity of calcareous earth mixed with such a soil, will immediately neutralize the acid, and destroy its powers: the soil, released from its baneful influence, will be rendered capable, for the first time, of exerting the fertility which it really possessed. The benefit thus produced is almost immediate: but though the soil will show a new vigour in its earliest vegetation, and may even double its first crop, yet no part of that increased product is due to the direct operation of the calcareous manure, but merely to the removal of acidity. The calcareous earth, in such a case, has not made the soil in the least richer,

[3] The antiseptic effect of vegetable acid in our soils receives some support from the facts established with regard to *peat soils*, in which vegetable acids have been discovered by chemical analysis: and though the peat or moss soils of Britain differ entirely from any soils in this country, still some facts relating to the former class, may throw light on the properties of our own soils, different as they may be. Not only does vegetable matter remain without putrefaction in peat soils and bogs, and serve to increase their depth by regular accessions from the annual growths, but even the bodies of beasts and men have been found unchanged under peat, many years after they had been covered. [William Aiton, *A Treatise on . . . Moss Earth . . .* (Air, 1811).] It is well known that the leaves of trees rot very quickly on the rich limestone soils of the western states, while the successive crops of several years growth may be always found on our acid woodland, in the different stages of their slow decomposition.

but has merely permitted it to bring into use the fertility that it had before, and which was concealed by the acid character of the soil. It will be a dangerous error for the farmer to suppose that calcareous earth can enrich soil by direct means. It destroys the worst foe of productiveness, and uses to the greatest advantage the fertilizing powers of other manures — but of itself it gives no fertility to soils, nor furnishes the least food to growing plants. These two kinds of action are by far the most powerful of the means possessed by calcareous earth, for fertilizing soils. It has another however of great importance — or rather two others, which may be best described together as the *power of altering the texture and absorbency of soils*.

At first it may seem impossible that the same manure could produce such opposite effects on soils, as to lessen the faults of being either too sandy, or too clayey — and the evils occasioned by both the want, and the excess of moisture. Contradictory as this may appear, it is strictly true as to calcareous earth. In common with clay, calcareous earth possesses the power of making sandy soils more close and firm — and in common with sand, the power of making clay soils lighter. When sand and clay thus alter the textures of soils, their operation is altogether mechanical; but calcareous earth must have some chemical action also, in producing such effects, as its power is far greater than either sand or clay. A very great quantity of clay would be required to stiffen a sandy soil perceptibly, and still more sand would be necessary to make a clay soil much lighter — so that the cost of such improvement would generally exceed the benefit obtained. Greater effects on the texture of soils are derived from less quantities of calcareous earth, in addition to the more valuable operation of its other powers.

Every substance that is open enough for air to enter, and the particles of which are not absolutely impenetrable, must absorb moisture from the atmosphere. Aluminous earth reduced to an impalpable powder, has strong absorbing powers. But this is not the form in which soils can act — and a close and solid clay will scarcely admit the passage of air or water, and therefore cannot absorb much moisture except by its surface. Through sandy soils, the air passes freely, but most of its particles are impenetrable by moisture, and therefore these soils are also extremely deficient in absorbent power. Calcareous earth, by rendering clay more open to the entrance of air, and closing partially the too open pores of sandy soils, increases the absorbent powers of both. To increase that power in any soil, is to enable it to draw supplies of moisture from the air, in the driest weather, and to resist more strongly the waste by evaporation, of light rains. A calcareous soil will so quickly absorb a hasty shower of rain, as to appear to have received less than adjoining land of different character: and yet if observed in summer when under tillage, some days after a rain, and when other adjacent land looks dry on the surface, the part made calcareous will still show the moisture remaining, by its darker colour. All the effects from this power of calcareous manures may be observed within a few years after their application — though none of them strongly marked, as they are on lands made calcareous by nature, and in which, time has aided and perfected the operation. These soils present great variety in their proportions of sand and clay — yet the most clayey is friable enough, and the most sandy, firm enough, to be considered soils of good texture: and they resist the extremes of both wet and dry seasons, better than any other soils whatever. Time, and the increase of vegetable matter, will bring those qualities to the same perfection, in soils made calcareous by artificial means.

The subsequent gradual accumulation of vegetable matter in soils to which calcareous manures have been applied, must also aid the improvement of their texture and absorbing power. The vegetable matter also darkens the colour of the soil, which makes it warmer by more freely absorbing the rays of the sun.

Additional and practical proofs of all the powers of calcareous earth will be furnished, when its use and effects as manure will be stated. I flatter myself however, that enough has already been said both to establish and account for the different capacities of soils for improvement by putrescent manures. If the power of fixing manures in soil, has been correctly ascribed to calcareous earth, that alone is enough to show that soils containing that ingredient in sufficient quantity, must become rich — that aluminous and silicious earths in any proportions, can never form other than a steril soil.

CHAPTER IX

THE PRACTICAL EFFECTS OF CALCAREOUS MANURES

PROPOSITION 5. *Calcareous manures will give to our worst soils a power of retaining putrescent manures, equal to that of the best — and will cause more productiveness, and yield more profit, than any other improvement practicable in Lower Virginia.*

THE theory of the constitution of fertile and barren soils, has been regularly discussed: it now remains to show its practical application in the use of calcareous earth as a manure. If the opinions which have been maintained are unsound, the attempt to reduce them to practice will surely expose their futility: and if they pass through that trial, agreeing with, and confirmed by facts, their truth and value must stand unquestioned. The belief in the most important of these opinions, (the incapacity of poor soils for improvement, and its cause,) directed the commencement of my use of calcareous manures; and the manner of my practice has also been directed entirely by the views which have been exhibited. Yet in every respect the results of practice have sustained the theory of the action of calcareous manures — unless there be found an exception in the damage which has been caused by applying too heavy dressings to weak lands.

My use of calcareous earth as manure, has been almost entirely confined to that form of it which is so abundant in the neighbourhood of our tide-waters — the beds of *fossil shells*, together with the earth with which they are found mixed. The shells are in various states — in some beds generally whole, and in others, reduced nearly to a coarse powder. The earth which fills their vacancies, and serves to make the whole a compact mass, in most cases is principally silicious sand, and contains no putrescent or valuable matter, other than the calcareous. The same effects might be expected from calcareous earth in any other form, whether chalk, limestone gravel, wood ashes, or lime — though the two last have other qualities besides the calcareous. During the short time that lime can remain *quick* or *caustic*, after being applied as manure, it exerts a solvent and decomposing power, sometimes beneficial and at others hurtful, which has no connexion with its subsequent and permanent action as calcareous earth.

These natural deposits of fossil shells are commonly, but very improperly, called *marl*. This misapplied term is particularly objectionable, because it induces erroneous views of this manure. Other earthy manures have long been used in England under the name of marl, and numerous publications have described their general effects, and recommended their use. When the same name is given here to a different manure, many persons will consider both operations as similar, and perhaps may refer to English authorities for the purpose of testing the truth of my opinions, and the results of my practice. But no two operations called by the same name, can well differ more. The process which it is my object to recommend, is simply the *application of calcareous earth in any form whatever, to soils wanting that ingredient,* and generally quite destitute of it — and the propriety of the application depends entirely on our knowing that the manure contains calcareous earth, and what proportion, and that the soil contains none. In England, the most scientific agriculturists apply the term *marl* correctly to a *calcareous clay*, of peculiar texture: but most authors, as well as mere cultivators, use it for any smooth soapy clay, which may, or may not, so far as they know, have any proportion of calcareous matter. Indeed, in most cases, they seem unconscious of the presence, as well as of the importance of that ingredient, by not alluding to it when attempting most carefully to point out the characters by which marl may be known. Still less do they inquire into deficiency of calcareous earth in soils proposed to be marled — but apply any earths which either science or ignorance may have called *marl*, to any soils within a convenient distance — and rely upon the subsequent effects to direct whether the operation shall be continued or abandoned. Authors of the highest character, (as Sinclair and Young,[1] for example,) when telling of the practical use, and valuable effects of marl, omit giving the strength of the manure, and generally even its nature — and in no instance have I found the ingredients of the soil stated, so that the reader might learn what operation really was described, or be enabled to form a judgment of its propriety. From all this, it follows that though what is called *marling* in England may sometimes (though very rarely as I infer) be the same chemical operation on the soil that I am recommending, yet it may also be, either applying clay to sand, or clay to chalk, or true marl to either of those soils — and the reader will generally be left to guess in every separate case, which of all these operations is meant by the term *marling*. For these reasons, the practical knowledge to be gathered from all this mass of written instruction on marling, will be far less abundant, than the inevitable errors and mistakes. The recommendations of marl by English authors, induced me very early to look at what was here called by the same name, as a means for improvement: but their descriptions of the manure convinced me that our marl was nothing like theirs, and thus actually deterred me from using it, until other views instructed me that its value did not depend on its having "a soapy feel," or on any mixture of clay whatever. [Appendix. F.]

But however incorrect and inconvenient the term may be, custom has too strongly fixed its application for any proposed

---

[1] [Sir John Sinclair (1754–1835), Scottish agricultural writer, and Arthur Young (1741–1820), noted English agricultural writer.]

change to be adopted. Therefore, I must submit to use the word *marl* to mean beds of fossil shells, notwithstanding my protest against the propriety of its being so applied.

The following experiments are reported, either on account of having been accurately made, and carefully observed, or as presenting such results as have been generally obtained on similar soils, from applications of fossil shells to nearly six hundred acres of Coggin's Point Farm. It has been my habit to make written memoranda of such things; and the material circumstances of these experiments were put in writing at the time they occurred, or not long after. Some of the experiments were, from their commencement, designed to be permanent, and their results to be measured as long as circumstances might permit. These were made with the utmost care. But generally, when precise amounts are not stated, the experiments were less carefully made, and their results reported by guess. Every measurement stated, of land, or of crop, was made in my presence. The average strength of the manure was ascertained by a sufficient number of analyses — and the quantity applied was known by measuring some of the loads, and having them dropped at certain distances. At the risk of being tedious, I shall state every circumstance supposed to affect the results of the experiments — and the manner of description, and of reference, necessary to use, will require a degree of attention that few readers may be disposed to give, to enable them to derive the full benefit of these details. When these operations were commenced, I knew of no other experiments having been made with fossil shells, except two, which had been tried long before, and were considered as proving the manure too worthless to be resorted to again.[2] My inexperience,

[2] The earliest of these old experiments was made at Spring Garden in Surry, about 1775. The extent marled was eight or ten acres, on poor sandy land. Nothing is known of the effects for the first twenty-five or thirty years, except that they were too inconsiderable to induce a repetition of the experiment. The system of cultivation was as exhausting as was usual during that time. Since 1812, the farm has been under mild and improving management generally. No care has been taken to observe the progress of

and total want of any practical guide, caused my applications, for the first few years, to be frequently injudicious, particularly as to the quantities laid on. For this reason, these experiments show what was actually done, and the effects thence derived, and not what better information would have directed, as the most profitable course.

The measurements of corn that will be reported were all made at the time and place of gathering. The measure used for all except very small quantities, was a barrel holding five bushels when filled level, and which being filled twice with ears of corn, well shaken to settle them, and heaped, was estimated to make five bushels of grain — and the products will be reported in *grain*, according to this estimate. This mode of measurement will best serve for comparing results — but in most cases it is far from giving correctly the actual quantity of dry and sound grain, for the following reasons. The common large soft grained white corn was the kind cultivated, and which was always cut down for sowing wheat before the best matured was dry enough to grind, or to put

---

either improvement or exhaustion on the marled piece: but there is no doubt but that the product has continued for the last fifteen years better than that of the adjacent land. Mr. Francis Ruffin, the present owner of the farm, believed that the product was not much increased in favorable seasons; but when the other land suffered either from too much wet, or dry weather, the crop on the marled land was comparatively but slightly injured. The loose reports that have been obtained respecting this experiment, are at least conclusive in showing the permanency of the effects produced.

The other old experiment referred to, was made at Aberdeen, Prince George county, in 1803, by Mr. Thomas Cocke. Three small spots (neither exceeding thirty yards square,) of poor land, kept before and since generally under exhausting culture, were covered with this manure. He found a very inconsiderable early improvement, which he thought entirely an inadequate reward for the labour of applying the marl. The experiment being deemed of no value, was but little noticed until the commencement of my use of the same manure. On examination, the improvement appeared to have increased greatly on two of the pieces, but the third was evidently the worse for the application. For a number of years after making this experiment, Mr. Cocke considered it as giving full proof of the worthlessness of the manure. But more correct views of its mode of operation have since induced him to recommence its use, and no one has met with more success, or produced more valuable improvements.

up in cribs, and when the ears from the poorest land were in a state to lose considerably more by shrinking. Yet, for fear of some mistake occurring if measurements were delayed until the crop was gathered, these experiments were measured when the land was ploughed for wheat in October. The subsequent loss from shrinking would of course be greatest on the corn from the poorest and most backward land, as there, most defective and unripe ears would always be found. Besides, every ear, however imperfect or rotten, was included in the measurement. For these several reasons, the actual increase of product on the marled land was always greater than will appear from the comparison of quantities measured: and from the statements of all such early measurements, there ought to be allowed a deduction, varying from ten per cent. on the best and most forward corn, to thirty per cent. on the latest and most defective. Having stated the grounds of this estimate, practical men can draw such conclusions as their experience may direct, from the dates and amounts of the actual measurements that will be reported.

No grazing has been permitted on any land from which experiments will be reported, unless it is specially stated.

Experiments on recently cleared acid sandy soils.

As most of the experiments on new land were made on a single piece of twenty-six acres, a general description or plan of the whole will enable me to be better understood, by references to the annexed figure. It forms part of the ridge lying between James River and the nearest stream running into Powell's Creek. The surface is nearly level. The soil in its natural state very similar throughout, but the part next to the line B C somewhat more sandy, and more productive in corn, than the part next to A D — and in like manner, it is lighter along A *e*, than nearer to D *f*. The whole soil, a grey silicious acid loam, not more than two inches deep at first, resting on

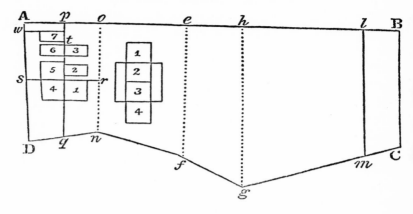

a yellowish sandy subsoil from one to two feet deep, when it changes to clay. Natural growth mostly pine — next in quantity, oaks of different kinds — a little of dogwood and chinquapin — whortleberry bushes throughout in plenty. The quality of the soil is better than the average of ridge lands in general.

EXPERIMENT 1.

The part B C *g h*, about eleven acres, grubbed and cut down in the winter of 1814–15 — suffered to lie three years with most of the wood and brush on it. February 1818, my earliest application of marl was made on B C *m l*, about 2½ acres. Marl, $^{33}/_{100}$ of calcareous earth, and the balance silicious sand, except a very small proportion of clay: the shelly matter finely divided. Quantity of marl to the acre, one hundred and twenty-five to two hundred heaped bushels. The whole B C *g h* coultered, and planted in its first crop of corn.

Results. 1818. The corn on the marled land, evidently much better — supposed difference, forty per cent.

1819. In wheat. The difference as great, perhaps more so — particularly to be remarked from the commencement to the end of winter, by the marled part preserving a green colour, while the remainder was seldom visible from a short distance, and by the spring, stood much thinner, from the greater number of plants having been killed. The line of separation very perceptible through both crops.

1820. At rest. During the summer marled all B C *g h*, at the rate of five hundred bushels, without excepting the space before covered, and a small part of that made as heavy as one thousand bushels, counting both dressings. The shells now generally coarse — average strength of the marl, $3\frac{7}{100}$ of calcareous earth. In the winter after, ploughed three inches deep as nearly as could be, which made the whole new surface yellow, by bringing barren subsoil to the top.

Results continued. 1821. In corn. The whole a remarkable growth for such a soil. The oldest (and heaviest) marled piece better than the other, but not enough so to show the dividing line. The average product of the whole supposed to have been fully twenty-five bushels to the acre.

1822. In wheat — and red clover sowed on all the old marling and one or two acres adjoining. A severe drought in June killed the greater part of the clover, but left it much the thickest on the oldest marled piece, so as again to show the dividing line, and to yield in 1823, two middling crops to the scythe — the first that I had known obtained from any acid soil, without high improvement from putrescent manures.

1823. At rest — nothing taken off, except the clover on B C *m l*.

1824. In corn — product seemed as before, and its rate may be inferred from the actual measurements on other parts, which will be stated in the next experiment, the whole being now cleared, and brought under like cultivation.

EXPERIMENT 2.

The part *e f n o*, cleared and cultivated in corn at the same times as the preceding — but treated differently in some other

s had been deprived of nearly all its wood, and
rnt, at the time of cutting down — and its first
1 (1818) being very inferior, was not followed
1819. This gave two years of rest before the crop
nd five years rest out of six, since the piece had
been cu̲ ̲ ̲wn. As before stated, the soil rather lighter on the
side next to *o e*, than *n f*.

March 1821. A measured acre near the middle, covered
with six hundred bushels of calcareous sand ($^{20}/_{100}$,) the upper layer of another body of fossil shells.

Results. 1821. In corn. October — the four adjoining
quarter acres, marked 1, 2, 3, 4, extending nearly across the
piece, two of them within, and two without the marled part,
measured as follows:

No. 1. not marled 6⅛ ⎫	average to the acre	22½
No. 4. - - - 5⅛ ⎭	bushels of grain.	
No. 2. marled - 8½ ⎫	average	33¼
No. 3. - - - 8⅛ ⎭	bushels.	

The remainder of this piece was marled before sowing
wheat in 1821.

1823. At rest.

1824. In corn — distance 5½ x 3¼ feet, making 2436 stalks
to the acre. October 11th measured two quarter acres very
nearly coinciding with Nos. 2 and 3 in the last measurement.
They now made

No.2. 7 bushels 3¼ pecks — or		⎫
per acre - - - - - - 31.1		⎬ average 31.2½
No. 3. 8 bushels, - - - - 32		⎭
Average in 1821, - - - -		33.1

EXPERIMENT 3.

The part *e f g h* was cut down in January 1821, and the
land planted in corn the same year. The coultering and after-

tillage very badly executed, on account of the number of whortleberry and other roots. As much as was convenient was marled at six hundred bushels ($^{37}/_{100}$) and the dressing limited by a straight line. Distance of corn 5½ x 3½ feet — 2262 stalks to the acre.

Results. 1821. October — On each side of the dividing line, a piece of twenty-eight by twenty-one corn hills measured as follows:

No. 1. 588 stalks, not marled, 2 bushels,
equal to   7¾ the acre,
   No. 2. 588   marled,   4¼   16⅝

1822. In wheat, the remainder having been previously marled.

1823. At rest. During the following winter it was covered with a second dressing of marl at 250 bushels ($^{45}/_{100}$) making 850 bushels to the acre, altogether.

1824. In corn. Two quarter acres, chosen as nearly as possible on the same spaces that were measured in 1821, produced as follows:

No. 1.   8   bushels, 2 pecks, or to the acre,   34
   The same in 1821, before marling,   7.3¼

No. 2. 7 bushels, 2½ pecks, or to the acre,   30.2
   The same in 1821, after marling,   16.1½

1825. The whole twenty-six acres, including the subjects of all these experiments and observations, were in wheat. The first marled piece in Exp. 1, was decidedly the best — and a gradual delcine was to be seen to the latest. I have never measured the product of wheat from any experiment, on account of the great trouble and difficulty that would be encountered. Even if the wheat from small measured spaces could be reaped and secured separately, during the heavy labours of harvest, it would be scarcely possible afterwards to carry the different parcels through all the operations necessary to show exactly the clean grain derived from each. But without any separate measurement, all my observations convince me, that the increase of wheat from marling, is at least equal to that of corn, during the first few years — and is certainly greater afterwards, in comparison to its product before using marl.

It was from the heaviest marled part of Exp. 1, that soil was analyzed to find how much calcareous earth remained in 1826, (page 79.) Before that time the marl and soil had been well mixed by ploughing to the depth of five inches. One of the specimens of this soil then examined, consisted of the following parts — the surface and consequently the undecomposed weeds upon it, being excluded.

1000 grains of soil yielded

769	grains of silicious sand moderately fine,
15	finer sand,
784	
8	calcareous earth, from the manure applied,
180	finely divided clay, vegetable matter, &c.
28	loss in the process.
1000	

This part, it has been already stated, was originally lighter than the general texture of the land.

EXPERIMENT 4.

The four acres marked A D *n o* were cleared in the winter 1823-4. The lines *p q* and *r s* divide the piece nearly into quarters. The end nearest A *p o* is lighter, and best for corn, and was still better for the first crop, owing to nearly that half having been accidentally burnt over. After twice coultering, marl and putrescent manures were applied as follows; and the products measured, October 11th, the same year.

s q not marled nor manured — produced on a quarter acre
(No. 4.)   3   bushels
(soft and badly filled) or per acre,   12   bushels

q r and *r p;* marled at 800 bushels ($^{45}/_{100}$)
gave by three measurements of different
pieces —

Quarter acre (No. 1.) 5 bushels, very
nearly, or   19.3½
 Eighth (No. 2.) 2.3¼ ⎫ average ⎧ 22.2
 Eighth (No. 3.) 3.1¼ ⎭ 24.1½ ⎩ 27.
s t manured at 900 to 1100 bushels to the
acre, of which,

Quarter acre (No. 5.) with rotted corn-
stalks, from a winter cow-pen, gave
      5.2½   22.2
 Eighth acre (No. 6.) with stable manure, 35.2
 Eighth (No. 7.) covered with the
same heavy dressings of stable manure, and
of marl also, gave   4.2   36.
p w marled at 450 bushels, was not so
good as the adjoining *r p* at 800.

The distance was 5½ x 3¼ feet. Two of the quarter acres
were measured by the surveyor's chain (as were four other
of the experiments of 1824,) and found to vary so little from
the distance counted by corn rows, that the difference was
not worth notice.

1825. In wheat: the different marked pieces seemed to yield
in comparison to each other, proportions not perceptibly
different from those of the preceding crop — but the best
not equal to any of the land marled before 1822, as stated in
the 1st, 2nd, and 3rd experiments.

1827. Wheat, on a very rough and imperfect summer fallow. This was too exhausting a course (being a grain crop
added to the four shift rotation,) — but was considered necessary to check the growth of bushes that had sprung from the
roots still living. The crop was small, as might have been expected from its preparation.

1828. Corn — in rows five feet apart, and about three feet

of distance along the rows, the seed being dropped by the step. Owing to unfavourable weather, and to insects and other vermin, not more than half of the first planting of this field lived — and so much replanting of course caused its product to be much less matured than usual, on the weaker land. All the part not marled, (and more particularly that manured,) was so covered by sorrel, as to require ten times as much labour in weeding as the marled parts, which, as in every other case, bore no sorrel. October 15th, gathered and measured the corn from the following spaces, which were laid off (by the chain) as nearly as could be, on the same land as in 1824. The products of both years can be best compared by being presented in a tabular form.

MARK.	DESCRIPTION.	PRODUCT OF GRAIN PER ACRE.			
		1824. October 11.		1828. October 15.	
		Bush.	Pecks.	Bush.	Pecks.
s q	Not marled or manured,	12	0	21	1
q r	(No. 1.) Marled, at 800 bushels,	19	3½	28	1½
r p	(No. 2.) Marled, 800 bushels,	22	2 }	31	0¼
r p	(No. 3.) Marled, 800 bushels,	27			
s t	(No. 5.) Cow-pen manure,	22	2	25	2
s t	(No. 6.) Stable manure,	35	2	29	0
w t	(No. 7.) Marl and stable manure,	36	0	33	2
w p	Marled at 450 bushels,	Less than r p (800)		As much as r p (800)	

Experiments on acid clay soils, recently cleared.

The two next experiments were made on a field of thirty acres of very uniform quality, marled and cleared in 1826 and the succeeding years. The soil is very stiff, close, and intractable under cultivation — seems to contain scarcely any sand — but in fact, about one-half of it is composed of silicious sand, which is so fine, when separated, as to feel like flour. Only a small proportion of the sand is coarser than this state of impalpable powder. Aluminous earth of a dirty fawn colour forms nearly all of its remaining ingredients. Before clearing, the soil is not an inch deep, and all below for some feet is apparently composed of the like parts of clay and fine sand. This is decidedly the most worthless kind of soil, in its natural state, that our district furnishes. It is better for wheat than for corn, though its product is contemptible in every thing: it is difficult to be made wet, or dry — and therefore suffers more than other soils from both dry and wet seasons, but especially from the former. It is almost always either too wet or too dry for ploughing — and sometimes it will pass through both states, in two or three clear and warm days. If broken up early in winter, the soil, instead of being pulverized by frost, like most clay lands, runs together again by freezing and thawing — and by March, will have a sleek (though not a very even) crust upon the surface, quite too hard to plant in without another ploughing. The natural growth is principally white and red oaks, a smaller proportion of pine, and whortleberry bushes throughout.

EXPERIMENT 5.

On one side of this field a marked spot of thirty-five yards square was left out, when the adjoining land was marled at the rate of five hundred to six hundred bushels ($^{37}/_{100}$) to the acre. Paths for the carts were opened through the trees, and the marl dropped and spread in January 1826 — the land cleared the following winter. Most of the wood was carried off for fuel — the remaining logs and brush burnt on the ground, as usual, at such distances as were convenient to the labourers. This part was perhaps the poorer, because wood had previously been cut here for fuel, though only a few trees taken, here and there, without any thing like clearing the land.

Results. 1827. Planted in corn the whole recent clearing of fifteen acres — all marled, except the spot left out for experiment. Broken up late and badly, and worse tilled, as the land was generally too hard, until the season was too far advanced to save the crop. The whole product so small, that it was useless to attempt to measure the experiment. The difference would have been only between a few imperfect ears on the marled ground, and still less — indeed almost nothing — on that not marled.

1828. Again in corn: as well broken and cultivated as usual for such land. October 18th. Cut down four rows of corn running through the land not marled, and eight others, alongside on the marled — all fifty feet in length. The rows had been laid off for five and a half feet — were found to vary a few inches — for which the proper allowance was made by calculation. The spaces taken for measurement were caused to be so small, by a part of the corn having been inadvertently cut down and shocked, just before. The ears were shelled when gathered; and the products, measured in a vessel which held (by trial) $\frac{1}{80}$ of a bushel, were as follows:

On land not marled

4 rows, average 5 feet, and 50 in length, (500 square

    feet) - - - - - - - - - 13½ measures,

or to the acre - - - - - - - 7¼ bushels.

    On adjoining marled land

4 rows, average 5 feet 1½ inches × 50 feet = 512 square

    feet - - - - - - - - - 25¾ measures,

or to the acre - - - - - - - 13½ bushels.

4 next rows, 5 feet 4½ inches × 50 = 537 square

    feet - - - - - - - - - 27½ measures,

or to the acre - - - - - - - 14 bushels.

    1829. In wheat.

    1830. At rest — the weeds, a scanty cover.

    1831. In corn. October 20th. Measured by the chain equal spaces, and gathered and measured their products. The unmarled corn was so imperfectly filled, that it was necessary to shell it, for fairly measuring the quantity. The marled parcels, being of good ears generally, were measured as usual,

by allowing two heaped measures of ears, for one of grain.
On land not marled

363 square yards made - - - - - - 3 gallons,
    or to the acre - - - - - - - 5 bushels.
On marled land close adjoining on one side,
    363 square yards made rather more than 6 gallons
        — to the acre, - - - - - - 10 bushels
    363 square yards on another side, made not quite 8 gallons,
        or to the acre, - - - - - - 12 bushels.

The piece not marled coincided with that measured in
1828, as nearly as their difference of size and shape permitted
— as did the last named marled piece, with the two of 1828.
The last crop was greatly injured by the wettest summer that
I have ever known, which has caused the decrease of product
exhibited in this experiment — which will be best seen in this
form:

	PRODUCT OF GRAIN TO THE ACRE.	
	1828. October 18.	1831. October 20.
	Bush. Pecks.	Bush. Pecks.
Not marled, - - - - - - -	7 1	5 0
Marled, (averaged), - - - - -	13 3	11 0

EXPERIMENT 6.

The remainder of the thirty acres, was grubbed during the
winter 1826–7 — marled the next summer at five hundred
to six hundred bushels the acre: marl $^{4}\%_{100}$. A rectangle (A)
11 × 13 poles, was laid off by the chain and compass, and
left without marl. All the surrounding land supposed to be
equal in quality with A — and all level, except on the sides
E and B, which were partly sloping, but not otherwise different. The soil suited to the general description given before
— no natural difference known or suspected, between the
land on which Exp. 5 was made, and this, except that the

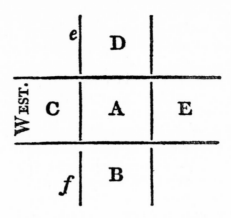

latter had not been robbed of any wood for fuel, before clear-ing. The large trees (all more than ten inches through,) were belted, and the smaller cut down in the beginning of 1828, and all the land west of the line *ef* was planted in corn. As usual, the tillage bad, and the crop very small. The balance, lying east of *ef*, was coultered once, but as more labour could not be spared, nothing more was done with it until the latter part of the winter 1829, when it was broken by two-horse ploughs, oats sowed and covered by trowel ploughs — then clover sowed, and a wooden-tooth harrow passed over to cover the seed, and to smooth down, in some measure, the masses of roots and clods.

Results. 1829. The oats produced badly — but yielded more for the labour required, than corn would have done. The young clover on the marled land was remarkably good, and covered the surface completely. In the unmarled part A, only two casts through had been sown, for comparison, as I knew it would be a waste of seed. This looked as badly as had been expected.

1830. The crop of clover would have been considered excellent for good land, and most extraordinary for so poor a soil as this. The strip sowed through A, had but little left alive, and that scarcely of a size to be observed, except one or two small tufts, where I supposed some marl had been deposited, by the cleaning of a plough, or that ashes had been left, from burning the brush. The growth of clover was left undisturbed until after midsummer, when it was grazed by my small stock of cattle, but not closely.

1831. Corn on the whole field. October 20th, measured carefully half an acre (10 × 8 poles) in A, the same in D, and half as much (10 × 4) in E. No more space could be taken on this side, for fear of getting within the injurious influence of the contiguous woods. No measurement was made on the side B, because a large oak, which belting had not killed, affected its product considerably. Another accidental circumstance prevented my being able to know the product of the side C — which however was evidently and greatly inferior to all the marled land on which oats and clover had been raised. This side had been in corn, followed by wheat, and then under its spontaneous growth of weeds. The corn on each of the measured spaces was cut down, and put in separate shocks — and on Nov. 25th, when well dried, the parcels were *shucked* and measured, without being moved. We had then been gathering and storing the crop, for more than fifteen days — and therefore these measurements may be considered as showing the amount of dry and firm grain, without any deduction being required for shrinkage.

		Bushels.	Pecks.
A (Half acre) made 7¼ bushels of ears, or of grain to the acre		7	1
D (Half acre) 16¾ - - - - -		16	3
E (Quarter acre) 11 - - - - -		22	

The sloping surface of the side E, prevented water from lying on it, and therefore it suffered less, perhaps not at all, from the extreme wetness of the summer, which evidently injured the growth on A and D, as well as of all the other level parts of the field.

EXPERIMENT 7.

Another piece of land of twenty-five acres, of soil and qualities similar to the last described (Exp. 5 and 6), was cleared in 1818, and about six acres marled in 1819, at about three hundred and fifty bushels. The course of cultivation was as follows.

1820 — Corn — benefit from marl very unequal — supposed to vary between twenty-five and eighty per cent.

1821. Wheat — the difference greater.

1822. At rest.

1823. Ploughed early for corn, but not planted. The whole marled at the rate of six hundred bushels ($^{40}\!/_{100}$,) again ploughed in August, and sowed in wheat in October. The old marled space, more lightly covered, so as to make the whole nearly equal.

1824. The wheat much improved.

1825 and 1826 — at rest.

1827. Corn.

1828. In wheat, and sowed in clover.

1829 — The crop of clover was heavier than any I had ever seen in this part of the country, except on rich natural soil, where gypsum was used, and acted well. The growth was thick, but unequal in height, (owing probably to unequal spreading of the marl,) — it stood from fifteen to twenty-four inches high. The first growth was mowed for hay, and the second left to improve the land.

1830. The clover not mowed. Fallowed in August, and sowed wheat in October, after a second ploughing.

1831. The wheat was excellent — almost heavy enough to be in danger of lodging. I supposed the product to be certainly twenty bushels — perhaps twenty-five, to the acre.

As it had not been designed to make any experiment on this land, the progress of improvement was not observed with much care. But whatever were the intermediate steps, it is certain that the land at first, was as poor as that forming the subjects of the two preceding experiments, in the unimproved state, (the measured products of which have been given) — and that its last crop was three or four times as great as could have been obtained, if marl had not been applied. The peculiar fitness of this kind of soil for clover after marling, will require further remarks, and will be again referred to hereafter.

CHAPTER X

THE EFFECTS OF CALCAREOUS MANURES, ON ACID SOILS REDUCED BY CULTIVATION

PROPOSITION 5. *Continued.*

MY use of fossil shells has been more extensive, on impoverished acid soils, than on all other kinds, and has never failed to produce striking improvement. Yet it has unfortunately happened, that the two experiments made on such land with most care, and on which I relied for evidence of the durable and increasing benefit from this manure, have had their effects almost destroyed, by the applications having been made too heavy. These experiments, like the 4th, 5th and 6th already reported, were designed to remain without any subsequent alteration, so that the measurement of their products once in every succeeding rotation, might exhibit the progress of improvement under all the different circumstances. As no danger was then feared from such a cause, marl was applied heavily, that no future addition might be required: and for this reason, I have to report my greatest disappointments, exactly in those cases where the most evident success and increasing benefits had been expected. However, these failures will be stated as fully as the most successful results — and they may at least serve to warn from the danger, if not to show the greatest profits of marling.

EXPERIMENT 8.

Of a poor silicious acid loam, seven acres were marled at the rate of only ninety bushels ($\frac{37}{100}$) to the acre: laid on and spread early in 1819.

Results. 1819. In corn — the benefit too small to be generally perceptible, but could be plainly distinguished along part of the outline, by comparing with the part not marled.

1820. Wheat — something better — and the effect continued to be visible on the weeds following, until the whole was more heavily marled in 1823.

EXPERIMENT 9.

In the same field, on soil as poor, and more sandy than the last described, four acres were marled at one hundred and eighty bushels ($^{37}\!/_{100}$), March 1819. A part of the same was also covered heavily with rotted barn-yard manure, which also extended through similar land not marled. This furnished for observation, land marled only — manured only — marled and manured — and some without either. The whole space, and more adjoining, had been manured five or six years before by summer cow-pens, and stable litter — of which no appearance remained after two years.

Results. 1819. In corn. The improvement from marl very evident — but not to be distinguished on the part covered also by manure, the effect of the latter so far exceeding that of the marl.

1820. In wheat. 1821 and 1822, at rest.

1823. In corn — 5½ × 3¼ feet — The following measurements were made on adjoining spaces on October 10th. The shape of the ground did not admit of larger pieces, equal in all respects, being measured, as no comparison of products had been contemplated at first, otherwise than by the eye.

From the part not marled —

				Bushels.	Quarts.
414 corn-hills made 75 quarts — or to the acre,				13	26
Marled only —					
414 - - - 100 - - - - -				18	12
Manured only —					
490 - - - 105 - - - - -				15	5

Marled and ma-
nured —

490 - - - 130 - - - - - 20   20

The growth on the part both marled and manured was evidently inferior to that of 1819: this was to be expected, as this small quantity of calcareous earth was not enough to fix half so much putrescent manure — and of course, the excess was as liable to waste, as if no marl had been used.

EXPERIMENT 10.

Twenty acres of sandy loam, on a sandy subsoil, covered in 1819 with marl of about $\frac{3}{100}$ average proportion of calcareous earth, and the balance silicious sand — at eight hundred bushels to the acre. This land had been long cleared, and much exhausted by cultivation: since 1813 not grazed, and had been in corn only once in four years, and as it was not worth sowing in wheat, had three years in each rotation to rest and improve by receiving all its scanty growth of weeds. The same course has been continued since 1819, except that wheat has regularly followed the crops of corn, leaving two years of rest, in four. This soil was lighter than the subject of any preceding experiment, except the ninth. On a high level part, surrounded by land apparently equal, a square of about an acre was staked off, and left without marl — which that year's work brought to two sides of the square.

Results. 1820. In corn: October 13th, three half acres of marled land were measured, and as many on that not marled, and close adjoining, and produced as follows:

	Not marled. Bushels.	Pecks.			Marled. Bushels.	Pecks.
Half acre A,	7	1	adjoining	C,	12	3
The same A,	7	1		D,	13	$3\frac{3}{4}$
Half acre B,	7	$2\frac{1}{2}$		E,	15	$0\frac{1}{2}$

The average increase being 12¾ bushels of grain to the acre: nearly 100 per cent. as measured, and more than 100, if the defective filling, and less matured state of the corn not marled, be considered. The whole would have lost more by shrinkage than is usual from equal products.

1821. The whole in wheat — much hurt by the wetness of the season. The marled part more than twice as good as that left out.

1822 and 1823. At rest. A good cover of carrot weeds and other kinds had succeeded the former growth of poverty grass and sorrel, and every appearance promised additional increase to the next cultivated crop. Nov. 1823, when the next ploughing was commenced, the soil was found to be evidently deeper, of a darker colour, and firmer, yet more friable. The two-horse ploughs with difficulty (increased by the cover of weeds,) could cut the required depth of five inches, and the slice crumbled as it fell from the mould-board. But as the furrows passed into the part not marled, an imme- diate change was seen, and even *felt* by the ploughman, as the cutting was so much more easy, that care was necessary to prevent the plough running too deep — and the slices turned over in flakes, smooth and sleek from the mould-board, like land too wet for ploughing, which however was not the case. The marling of the field was completed, at the same rate, (eight hundred bushels,) which closed a third side of the marked square. The fourth side was my neighbour's field.

1824. In corn. The newly marled part showed as early and as great benefit as was found in 1820 — but was very inferior to the old, until the latter was ten or twelve inches high, when it began to give evidence of the fatal effects of using this manure too heavily. The disease thus produced became worse and worse, until many of the plants had been killed, and still more were so stunted, as to leave no hope of their being otherwise than barren. The effects will be known from the measurements, which were made nearly on the same ground as the corresponding marks in 1820, and will be exhibited in the table, together with the products of the succeeding rotation. Besides the general injury suffered here in 1824, there were one hundred and three corn hills in one of the measured quarter acres (in C) or more than one-sixth, entirely barren, and eighty-nine corn hills in another quarter acre (D.) In counting these, none of the missing hills were included, as these plants might have perished from other causes.

		PRODUCT IN SHELLED CORN, PER ACRE.		
		1820. October 13.	1824. October 16.	1828. October 13.
		Bush. Pecks.	Bush. Pecks.	Bush. Pecks.
A	Not marled,	14 2	16 1	11 3½
B	Not marled until 1823,	15 1	28 0	19 2
C	} Marled,	25 0	19 2	15 0
D		27 3½	20 0	19 0
E		30 1		

The crops of wheat were less injured than the corn.

For the crop of 1828, ploughed with three mules to each plough, from six to seven inches deep — seldom turning up any subsoil, (which formerly was within three inches of the surface,) and the soil looking still darker and richer than before. The ploughing of the square not marled, no where exceeded six inches.

EXPERIMENT 11.

The ground on which this experiment was made, was in the midst of nineteen or twenty acres of soil apparently similar in all respects — level, grey sandy loam, cleared about thirty years before, and reduced as low by cultivation as such soil could well be. The land that was marled and measured was about two hundred yards distant from the second experiment, and both places are supposed to have been originally similar in all respects. This land had not been cultivated since

1815, when it was in corn — but had been once ploughed since, in Nov. 1817, which had prevented broom grass from taking possession. The ploughing then was four inches deep, and in five and a half feet beds, as recommended in *Arator*. The growth in the year 1820, presented but little except poverty grass, running blackberry, and sorrel — and the land seemed very little if at all improved by its five successive years of rest. A small part of this land was covered with calcareous sand $^{20}/_{100}$ — quantity not observed particularly, but probably about six hundred bushels.

Results. 1821 — Ploughed level, and planted in corn — distance 5½ × 3½ feet. The measurement of spaces nearly adjoining, made in October, was as follows:

23 × 25 corn hills, not marled, made 2⅛
     bushels, or   very
      per acre, - - - - - -   8⅜ ⎬ nearly.
23 × 25 corn hills, marled   5⅝ 22⅛

1822. At rest. Marled the whole, except a marked square of fifty yards, containing the space measured the preceding year. Marl $^{45}/_{100}$ and finely divided — three hundred and fifty bushels to the acre — from the same bed as that used for experiment 4th. In August, ploughed the land, and sowed wheat early in October.

1823. Much injury sustained by the wheat from Hessian fly, and the growth was not only mean, but very irregular — but it was supposed that the first marled place was from fifty to one hundred per cent. better than the last, and the last superior to the included square not marled, in as great a proportion.

1824. Again in corn. The effects of disease from marling were as injurious here, both on the new and old part, as those described in Experiment 10. No measurement of products made, owing to my absence, when the corn was cut down for sowing wheat.

1825. The injury from disease less on the wheat, than on

the corn of the last year on the latest marling, and none perceptible on the oldest application. This scourging rotation of three grain crops in four years, was particularly improper on marled land, and the more so on account of its poverty.

1826. White clover had been sown thickly over forty-five acres, including this part, on the wheat, in January 1825.

In the spring of 1826, it formed a beautiful green though low cover to even the poorest of the marled land. Marked spots, which were so diseased by over-marling, as not to produce a grain of corn or wheat, produced clover at least as good as other places not injured by that cause. The square, which had been sowed in the same manner, and on which the plants came up well, had none remaining by April 1826, except on a few small spots, all of which together would not have made three feet square. The piece not marled, white with poverty grass, might be seen, and its outlines traced at some distance by its strong contrast with the surrounding dark weeds in winter, or the verdant turf of white clover the spring before.

1827. Still at rest. No grazing allowed on the white clover.

1828. In corn — the land broken in January, five inches deep. October 14th made the following measurements.

In the square not marled 105 × 104½ feet (thirty-six square yards more than a quarter of an acre,) made one barrel of ears — or of grain, to the acre,

	Bushels.	Pecks.
	9	1 ¾
The same in 1821,	8	1 ½
Gain,	1	0 ¼
Old marling — 105 × 104½ feet — 2 ¼ barrels,	22	2
The same in 1821,	22	0 ½
Gain,		1 ½

New marling, 105 × 104½ feet, on the side that seemed to be the most diseased, 1⅓ barrels — or nearly twelve bushels.

EXPERIMENT 12.

On nine acres of sandy loam, marled in 1829 at four hundred bushels ($^{25}/_{100}$) nearly an acre was manured during the same summer, by penning cattle: with the expectation of preserving the manure, double the quantity of marl, eight hundred bushels in all, was laid on that part. The field in corn in 1820 — in wheat, 1821 — and at rest 1822 and 1823.

Result. 1824. In corn, the second rotation after marling. The effects of the dung has not much diminished, and that part shows no damage from the quantity of marl, though the surrounding corn, marled only half as thickly, gave signs of general, though very slight injury from that cause.

EXPERIMENT 13.

Nearly two acres of loamy sand, was covered with farm-yard manure, and marl ($^{45}/_{100}$) at the same time, in the spring of 1822, and tended in corn the same year, followed by wheat. The quantity of marl not remembered — but it must have been heavy (say not less than six hundred bushels to the acre) as it was put on to fix and retain the manure, and I had then no fear of damage from heavy dressings.

Result. 1825. Again in corn — and except on a small spot of sand almost pure, no signs of disease from over-marling.

EXPERIMENT 14.

The soil known in this part of the country by the name of "free light land" has so peculiar a character, that it deserves a particular notice. It belongs to the slopes and waving lands, between the ridges and the water courses, but has nothing of the durability, which slopes of medium fertility sometimes possess. In its woodland state it would be called rich, and may remain productive for a few crops after clearing — but it is rapidly exhausted, and when poor, seems as unimprovable by

vegetable manures as the poorest ridge lands. In its virgin
state, this soil might be supposed to deserve the name of
neutral — but its productive power is so fleeting, and acid
growths and qualities so surely follow its exhaustion, that we
must infer that it is truly an acid soil.

The subject of this experiment presents soil of this kind
with its peculiar characters unusually well marked. It is a
loamy sandy soil (the sand coarse) on a similar subsoil of
considerable depth. The surface waving — almost hilly in
some parts. The original growth principally red oak, hickory,
and dogwood — not many pines, and very little whortleberry. Cut down in 1816 and put in corn the next year. The
crop was supposed to be twenty-five bushels to the acre.
Wheat succeeded, and was still a better crop for so sandy a
soil — making twelve to fifteen bushels, as it appeared standing. After a year of rest, and not grazed, the next corn crop
of 1820, was evidently considerably inferior to the first — and
the wheat of 1821, (which however was a very bad crop
from too wet a season) could not have been more than five
bushels to the acre. In January 1820, a piece of 1½ acres
was limed, at one hundred bushels the acre. The lime being
caught by rain before it was spread, formed small lumps of
mortar on the land, and produced no benefit on the corn
of that year, but could be seen slightly in the wheat of 1821.
The land again at rest in 1822 and 23, when it was marled,
at six hundred bushels, ($^{37}\!/_{100}$) without omitting the limed
piece — and all sowed in wheat that fall. In 1824, the wheat
was found to be improved by the marl, but neither that, nor
the next, of 1828, was equal to its earliest product of wheat.
The limed part showed injury from the quantity of manure,
in 1824, but none since. The field was now under the regular
four shift rotation, and continued to recover — but did not
surpass its first crop until 1831, when it brought rather more
than thirty bushels of corn to the acre — five or six more
than its supposed first crop.

Adjoining this piece, six acres of similar soil were grubbed and belted in August 1826 — marl at six hundred to seven hundred bushels ($^{37}/_{100}$) spread just before. But few of the trees died until the summer of 1827. 1828, planted in corn: the crop did not appear heavier than would have been expected if no marl had been applied — but no part had been left without, for comparison. 1829 — wheat. 1830, at rest. 1831, in corn, and the product supposed to be near or quite thirty-five bushels — or an increase of thirty-five or forty per cent. on the first crop. No measurement was made — but the product was estimated by comparison with an adjacent piece, which measured thirty-one bushels, and which was evidently something inferior to this piece.[1]

The operation of marl on this kind of soil, seems to add to the previous product very slowly, compared with other soils — but it is not the less effectual and profitable, in fixing and retaining the vegetable matter accumulated by nature, which otherwise would be quickly dissipated by cultivation, and lost forever.

[1] 1000 grains of this soil, taken in 1826 from the part that had been both limed and marled, was found to consist of
    811 of silicious sand moderately coarse, mixed with a few grains of coarse shelly matter.
    158 — finely divided earthy matter, &c.
     31 — loss.
     ―――
    1000

At the same time, from the edge of the adjoining woodland which formed the next described clearing, and which had not then been marled, a specimen of soil was taken from between the depths of one and three inches — and found to consist of the following proportions. This spot was believed to be rather lighter than the other in its natural state.
    865 grains of silicious sand, principally coarse —
    107 — finely divided earthy matter, &c.
     28 — loss.
    ―――
    1000

CHAPTER XI

EFFECTS OF CALCAREOUS MANURES ON EX-
HAUSTED ACID SOILS, UNDER THEIR
SECOND GROWTH OF TREES

PROPOSITION 5. *Continued.*

NOT having owned much land under a second growth of pines, I can only refer to two experiments of this kind. The improvement in both these cases has been so remarkable, as to induce the belief that the old fields to be found on every farm, which have been exhausted and turned out of cultivation thirty or forty years, offer the most profitable subjects for the application of calcareous manures.

EXPERIMENT 15.

May 1826. Marled about eight acres of land under its second growth, by opening paths for the carts, ten yards apart. Marl $^{40}\!/_{100}$, put five hundred to six hundred bushels to the acre — and spread in the course of the summer. In August, belted slightly all the pines that were as much as eight inches through, and cut down or grubbed the smaller growth, of which there was very little. The pines (which were the only trees,) stood thick, and were mostly from eight to twelve inches in diameter — eighteen inches where standing thin. The land joined Exp. 14 on one side — but this is level, and on the other side joins ridge woodland, which soon becomes like the soil of Exp. 1. This piece, in its virgin state, was probably of a nature between those two soils — but more like the ridge soil, than the "free light land." No information has been obtained as to the state of this land when its former cultivation was abandoned. The soil, (that is what has since been turned by the plough,) a whitish loamy sand, on a subsoil of the same: in fact, all was subsoil, before ploughing, except half an inch or three quarters, on the top, which was principally composed of rotted pine leaves. Above this thin layer, were the later dropped and unrotted leaves, lying loosely several inches thick.

The pines showed no symptom of being killed, until the autumn of 1827, when their leaves began to have a tinge of yellow. To suit the cultivation with the surrounding land, this piece was laid down in wheat for its first crop, in October, 1827. For this purpose the few logs, the boughs, and grubbed bushes were heaped, but not burnt — the grain sowed on the coat of pine leaves, and ploughed in by two-horse ploughs, in as slovenly a manner as may be supposed — and a wooden-tooth harrow passed over to pull down the heaps of leaves, and roughest furrows.

Results. The wheat was thin, but otherwise looked well while young. The surface was again soon covered, by the leaves dropping from the now dying trees. On April 2d 1828, most of the trees were nearly dead, though but few of them entirely. The wheat was then taller than any in my crop — and when ripe, was a surprising crop, for such land, and such tillage.

1829 and 1830. At rest. Late in the spring of 1830 an accidental fire passed over the land — but the then growing vegetation prevented all the older cover being burnt, though some was destroyed every where.

1831. In corn. The growth excited the admiration of all who saw it, and no one estimated the product so low as it actually proved to be. A square of four (two pole) chains, or four-tenths of an acre, measured on November 25th, yielded at the rate of thirty-one and three-eighths bushels of grain to the acre.

EXPERIMENT 16.

In a field of acid sandy loam, long under the usual cultivation, a piece of five or six acres was covered by a second growth of pines thirty-nine years old, as supposed from that number of rings being counted on some of the stumps. The largest trees were eighteen or twenty inches through. This ground was altogether on the side of a slope, steep enough to lose soil by washing, and more than one old shallow gulley remained to confirm the belief of the injury that had been formerly sustained from that cause. These circumstances, added to all the surrounding land having been continued under cultivation, made it evident that this piece had been turned out of cultivation because greatly injured by tillage. It was again cut down in the winter of 1824–5 — many of the trees furnished fence rails, and fuel — and the remaining bodies were heaped and burnt some months after, as well as the large brush. In August it was marled, supposed at six-hundred bushels ($\frac{37}{100}$) — twice coultered in August and September, and sowed in wheat — the seed covered by trowel ploughs. The leaves and much of the smaller brush left on the ground, made the ploughing troublesome and imperfect. The crop (1826) was remarkably good — and still better were the crops of corn and wheat in the ensuing rotation, after two years of rest. On the last crop of wheat (1830) clover was sowed — and mowed for hay in 1831. The growth stood about eighteen inches high, and never have I seen so heavy a crop on sandy and acid soil, even from the heaviest dunging, the utmost care, and the most favourable season. The clover grew well in the bottoms of the old gullies, which are still plainly to be seen, and which no means had been used to improve, except what all the land had received.

CHAPTER XII

EFFECTS OF CALCAREOUS MANURES, ON
NEUTRAL SOILS, ALONE, OR
WITH GYPSUM

PROPOSITION 5. *Continued.*

APPLICATIONS of calcareous earth alone, to calcareous soils, are so manifestly useless, that only two experiments of that kind have been made, neither of which has had any improving effect that could be observed, in the twelve years that have since elapsed.

When calcareous manures have been applied to neutral soils, whether new or worn, no perceptible benefit has been obtained on the earliest crops. The subsequent improvement has gradually increased as would be expected from the power of fixing manures, attributed to calcareous earth. But however satisfactory these general results are to myself, they are not such as could be reported in detail, with any advantage to other persons. It is sufficiently difficult to make fair and accurate experiments, where early and remarkable results are expected. But no cultivator of a farm can bestow enough care, and patient observation, to obtain true results from experiments that scarcely will show their first feeble effects, in several years after the commencement. On a mere experimental farm, such things may be possible — but not where the main object of the farmer is profit from his general and varied operations. The effects of changes of season, of crops, of the mode of tillage — the auxiliary effects of other manures — and many other circumstances, would serve to defeat any observations of the progress of a slow improvement, though the ultimate result of the general practice might be abundantly evident.

Another cause of my being unable to state with any precision the practical benefit of marling neutral soils, arises from the circumstance that nearly all the calcareous manure thus applied, has been accompanied by a natural admixture of gypsum: and though I feel confident in ascribing some effects to one, and some to the other of these two kinds of manure, yet this division of operation must rest merely on opinion, and cannot be received as certain, by any other than him who makes and carefully observes the experiments. Some of these applications will be described, that other persons may draw their own conclusions from them.

The cause of these manures being applied in conjunction was this. A singular bed of marl lying under Coggin's Point, and the only one within a convenient distance to most of the neutral soil, contains a very small proportion (perhaps one to two per cent.) of gypsum,[1] scattered irregularly through

---

[1] What led me to suspect the presence of gypsum in this bed of fossil shells, was the circumstance that throughout its whole extent of near a mile along the river bank, this bed lies on another earth, of peculiar character and appearance, and which in many places exhibits gypsum, in crystals of various sizes. This earth has evidently once been a bed of fossil shells, like what still remains above — but nothing now is left of the shells, except numerous impressions of their forms. Not the smallest proportion of calcareous earth can be found — and the gypsum into which it must have been changed (by meeting with sulphuric acid, or sulphuret of iron,) has also disappeared in most places, and in others, remains only in small quantities — say from the smallest perceptible proportion, to fifteen or twenty per cent. of the mixed mass. In some rare cases, this gypseous earth is sufficiently abundant to be used profitably as manure, as has been done, by Mr. Thomas Cocke of Tarbay, as well as myself. It is found in the greatest quantity, and also the richest in gypsum, at Evergreen, two miles below City Point. There the gypsum frequently forms large crystals of varied and beautiful forms. The distance that this bed of gypseous earth extends is about seven miles, interrupted only by some formations of soil by alluvion.

In the bed of gypseous marl above described, there are regular layers of a calcareous rock, which was too hard to use profitably for manure, and which caused the greatest impediment to obtaining the softer part. This rock contains between eighty-five and ninety per cent. of pure calcareous earth, besides a little gypsum and iron. It makes excellent lime for cement, mixed with twice its bulk of sand — and has been used for part of the brick-work,

the mass, seldom visible, though sometimes to be met with in small crystals. The calcareous ingredient is generally about $\frac{53}{100}$ — sometimes $\frac{60}{100}$. If this manure had been used before its *gypseous* quality was discovered, all its effects would have been ascribed to calcareous earth alone, and the most erroneous opinions might thence have been formed of its mode of operation.

This gypseous marl has been used on fifty-six acres, most of which was neutral soil — and generally, if not universally, with early as well as permanent benefits. The following experiments show results more striking than have been usually obtained, but all agree in their general character.

EXPERIMENT 17.

1819. Across the shelly island numbered 3 in the examinations of soils, (page 55) but where the land was less calcareous, a strip of three quarters of an acre was covered with muscle-shell marl. Touching this through its whole length, another strip was covered with gypseous marl, ($\frac{53}{100}$) at the rate of two hundred and fifty bushels.

Results. 1819. In corn. No perceptible effect from the muscle shells. The gypseous marling considerably better than on either side of it.

1820. Wheat — less difference.

1821. Grazed. Natural growth of white clover thickly set

---

and all the plastering of my present dwelling house, and for several of my neighbours' houses. The whole body of marl also contains a minute proportion of some soluble salts, which possibly may have some influence on the operation of the substance, as manure, or cement.

Thus, from the examination of a single body of marl, there have been obtained not only a rich calcareous manure, but also gypsum, and a valuable cement. Similar formations may perhaps be abundant elsewhere, and their value unsuspected, and likely to remain useless. This particular body of marl has no outward appearance of even its calcareous character. It would be considered, on slight observation, a mass of gritty clay, of no worth whatever.

on the gypseous marling, much thinner on the muscle shells, and still less of it where no marl had been applied.

The whole field afterwards was put in wheat on summer fallow every second year, and grazed the intervening year — a course very unfavourable for observing, or permitting to take place, any effects of gypsum. Nothing more was noted of this experiment until 1825, when cattle were not turned in until the clover reached its full size. The strip covered with gypseous marl showed a remarkable superiority over the other marled piece, as well as the land which was still more calcareous by nature, and produced better in 1820. In several places, the white clover stood thickly a foot in height.

EXPERIMENT 18.

A strip of a quarter acre passing through rich black neutral loam, covered with gypseous marl at two hundred and fifty bushels.

Results. 1818. In corn. By July, the marled part seemed the best by fifty per cent., but afterwards the other land gained on it, and little or no difference was apparent, when the crop was matured.

1819. Wheat — no difference.

1820 and 1821. At rest. In the last summer, the marled strip could again be easily traced, by the entire absence of sorrel, (which had been gradually increasing on this land since it had been secured from grazing,) and still more by its very luxuriant growth of bindfoot clover, which was thrice as good as that on the adjoining ground.

EXPERIMENT 19.

1822. On a body of neutral soil which had been reduced quite low, but was well manured in 1819 when last cultivated, gypseous marl was spread on nine acres, at the rate of three

hundred bushels. This terminated on one side at a strip of
muscle-shell marl ten yards wide — its rate not remembered,
but it was certainly thicker in proportion to the calcareous
earth contained, than the other, which I always avoided laying
on heavily, for fear of causing injury by too much gypsum.
The line of division between the two marls, was through a clay
loam. The subsoil was a retentive clay, which caused the rain
water to keep the land very wet through the winter, and early
part of spring.

Results. 1822. In corn, followed by wheat in 1823: not
particularly noticed — but the benefits must have been very
inconsiderable. All the muscle-shell marling, and four acres
of the gypseous, sowed in red clover, which stood well, but
was severely checked, and much of it killed, by a drought in
June, when the sheltering wheat was reaped. During the next
winter my horses had frequent access to this piece, and by
their trampling in its wet state, must have injured both land
and clover. From these disasters, the clover recovered surpris-
ingly, and in 1824, two mowings were obtained, which though
not heavy, were better than from any of my previous attempts
to raise this grass. In 1825, the growth was still better, and
yielded more to the scythe. This was the first time that I had
seen clover worth mowing on the third year after sowing —
and had never heard of its being comparable to the second
year's growth, any where in the lower country. The growth
on the muscle-shell marling was very inferior to the other,
and was not mowed at all the last year, being thin and low,
and almost eaten out by wire grass.

1826. In corn — and it was remarkable that the difference
shown the last year was reversed, the muscle-shell marling
now having much the best crop.

In these and other applications to neutral soils, I ascribe
the earliest effects entirely to gypsum, as well as the peculiar
benefit shown to clover, throughout. The later effects on
grain, are due to the calcareous earth in the manure.

Another opinion was formed from the effects of gypseous marl, which may lead to profits much more important than any to be derived from the limited use of this, or any similar mineral compound — viz: *that gypsum may be profitably used after calcareous manures, on soils on which it was totally inefficient before.* I do not present this as an established fact, of universal application — for the results of some of my own experiments are directly in opposition. But however it may be opposed by some facts, the greater weight of evidence furnished by my experiments and observations, decidedly supports this opinion. If correct, its importance to our low country is inferior only to the value of calcareous manures — which value, may be almost doubled, if the land is thereby fitted for the wonderful effects of gypsum on clover.

It is well known that gypsum has failed entirely as a manure on nearly all the land on which it has been tried in the tide-water district — and we may learn from various publications, that as little general success has been met with along the Atlantic coast, as far north as Long Island. To account for this general failure of a manure so efficacious elsewhere, some one offered a reason, which was received without examination, and which is still considered by many as sufficient, viz. that the influence of salt vapours destroyed the power of gypsum on and near the sea coast. But the same general worthlessness of that manure extends one hundred miles higher than the salt water of the rivers — and the lands where it is profitably used, are much more exposed to sea air. Such are the rich neutral soils of Carle's Neck, Shirley, Berkley, Brandon, and Sandy Point on James River, on all which gypsum on clover has been extensively and profitably used. On acid soils, I have never heard of enough benefit being obtained from gypsum to induce the cultivator to extend its use further than making a few small experiments. When any effect has been produced on an acid soil, (so far as instructed by my own experience, or the information of others,) it has been caused by applying to small spaces, comparatively large quantities — and even then, the effects were neither considerable, durable, nor profitable. Such have been the results of many small experiments made on my own acid soils — and very rarely was the least perceptible effect produced. Yet on some of the same soils, after marling, the most evident benefits have been obtained from gypsum on clover. The soils on which the 1st and 10th experiments were made, (at some distance from these experiments) had both been tried with gypsum, and at different rates of thickness, before marling, without the least effect. Several years after both had been marled, gypseous earth (from the bed described in the note to page 112,) was spread at twenty bushels the acre, which gave four bushels of pure gypsum, on clover, and produced in some parts, a growth I have never seen surpassed. It is proper to state that such results have been produced only by heavy dressings. Mr. Thomas Cocke of Tarbay has this last spring (1831) sowed nearly four tons of Nova-Scotia gypsum on clover on marled land, a continuation of the same ridge that my 1st, 2d, 3d, and 4th Experiments were made on, and very similar soil. His dressing, at a bushel to the acre, before the summer had passed, produced evident benefit, where it is absolutely certain that none could have been obtained before marling.

On soils naturally calcareous, I have in some experiments greatly promoted the growth of corn, by gypsum, and have doubled the growth of clover on my best land of that kind. When the marl containing gypsum was applied, benefit from that ingredient was almost certain to be obtained.

All these facts, if presented alone, would seem to prove clearly the correctness of the opinion, that the acidity of our soils caused the inefficacy of gypsum, and that the application of calcareous earth, which will remove the former, will also serve to bring the latter into useful operation. But this most desirable conclusion is opposed by the results of other ex-

periments, which though fewer in number, are as strong as
any of the facts that favour that conclusion. If the subject
was properly investigated, those facts apparently in opposition, might be explained so as no longer to contradict this
opinion — perhaps even help to confirm it. Good reasons,
deduced from established chemical truths, may be offered to
explain why the acidity of our soils should prevent the operation of gypsum: but it may be deemed premature to attempt
the explanation of any supposed fact, before every doubt of
its existence has been first removed. The subject well deserves
a more full investigation from those who can be aided by
more information, whether practical or scientific. [Appendix.
G.]

One of the circumstances will be mentioned, which appears
most strongly opposed to the opinion which has been advanced. On the poor acid clay soil, of such peculiar and base
qualities, which forms the subject of the 5th, 6th, and 7th
Experiments, gypsum has been sufficiently tried, and has produced not the least benefit, either before marling, or afterwards. Yet the growth of clover on this land after marling,
is fully equal to what might be expected from the best operation of gypsum. Now if it could be ascertained that a very
small proportion of either *sulphuric acid*, or the *sulphate of
iron* exists in this soil, it would completely explain away this
opposing fact, and make it the strongest support of my position. The sulphate of iron has been found in arable soil,[2] and
sulphuric acid has been detected in certain clays.[3] I have seen
on the same farm a clay of very similar appearance to this
soil, which had once contained one of these substances, as
was proved by the formation of crystallized sulphate of lime,
where the clay came in contact with calcareous earth. The
sulphate of lime was found in the small fissures of the clay,

---

[2] *Agr. Chem.*, p. 141.
[3] Kirwan, *Manures* [. . . *Applicable to the Various . . . Soils . . .* (London, 1796)].

extending sometimes one or two feet distant from the calcareous earth below. Precisely the same chemical change would
take place in a soil containing sulphuric acid, or sulphate of
iron, as soon as marl was applied. The sulphuric acid, (whether
free or combined with iron) would immediately unite with
the lime presented, and form gypsum, (sulphate of lime.)
Proportions of these substances too small perhaps to be detected by analysis, would be sufficient to form three or four
bushels of gypsum to the acre — more than enough to produce the greatest effect on clover — and to prevent any
benefit being derived from a subsequent application of gypsum.

CHAPTER XIII

THE DAMAGE CAUSED BY CALCAREOUS MANURE, AND ITS REMEDIES

PROPOSITION 5. *Continued.*

THE injury or disease in grain crops produced by marling has so lately been presented to our notice, that the collection and comparison of many additional facts will be required before its cause can be satisfactorily explained. But the facts already ascertained will show how to avoid the danger in future, and to find remedies for the evils already inflicted by the injudicious use of calcareous manures.

The earliest effect of this kind observed, was in May 1824, on the field containing Experiment 10. The corn on the land marled four years before, sprang up and grew with all the vigour and luxuriance that was expected from the appearances of increased fertility exhibited by the soil, as has already been described, (page 101.) About the 20th of May the change commenced, and the worst symptoms of the disease were seen by the 11th of June. From having as deep a colour as young corn shows on the richest soils, it became of a pale sickly green. The leaves, when closely examined, seemed almost transparent — afterwards were marked through their whole length by streaks of rusty red, separated very regularly by what was then more of yellow than green — and next began to shrivel, and die downwards from their extremities. The growth of many of the plants was nearly stopped. Still some few showed no sign of injury, and maintained the vigorous growth which they began with, so as by contrast more strongly to mark the general loss sustained. The appearance of the field was such, that a stranger would have supposed that he saw the crop of a rich soil exposed to the worst ravages of some destructive kind of insect: but neither on the roots or stalks of the corn could any thing be found to support that opinion. Before the 1st of August, this gloomy prospect had improved. Most of the plants seemed to have been relieved of the infliction, and to grow again with renewed vigour. But before that time, many were dead, and it was impossible that the others could so fully recover as to produce any thing approaching a full crop for the land. It has been shown in the report of the products of Exp. 10, what diminution of crop was then sustained — and that the evil was not abated by the next cultivation. Still, neither of the diseased measured pieces has fallen as low as to its product before marling — nor do I think that such has been the result on any one acre on my farm, though many smaller spots have been rendered incapable of yielding a grain of corn or wheat.

The injury caused to wheat by marling is not so easy to describe, though abundantly evident to the observer. Its earliest growth, like that of corn, is not affected. About the time for heading, the plants most diseased appear as if they were scorched, and when ripe, will be found very deficient in grain. On very poor spots, from which nearly all the soil has been washed, sometimes fifty heads of wheat taken together would not furnish as many grains of wheat. This crop, however, suffers less than corn on the same land — perhaps because its growth is nearly completed by the time that the season begins, to which the ill effects of calcareous manures seem confined.

When these unpleasant discoveries were first made, two hundred and fifty acres had already been marled so heavily, that the same evil was to be expected to visit the whole. My labours thus bestowed for years had been greatly and unnecessarily increased — and the excess, worse than being thrown away, has served to take away that increase of crop, that lighter marling would have ensured. But though much

and general injury was afterwards sustained from the previous
work, yet it was lessened, and sometimes entirely avoided,
by the remedial measures which were adopted. My observation and comparison of all the facts presented, led to the
following conclusions, and pointed out the course to avoid
the recurrence of the evil, and the means to lessen or remove
it, where it had already been inflicted.

1st. No injury has been sustained on any soil of my farm
by marling not more heavily than two hundred and fifty
heaped bushels to the acre, with marl of strength not exceeding $^{4}\%_{00}$ of calcareous earth.

2d. Dressings twice as heavy seldom produce damage to
the first crop on any soil — and never on the after crops on
any calcareous, or good neutral soil — nor on any acid soil
supplied plentifully with vegetable matter.

3d. On acid soils marled too heavily, the injury is in proportion to the extent of one or all these circumstances of the
soil — poverty, sandiness, and previous severe cropping and
grazing.

4th. Clover, both red and white, will live and flourish on
the spots most injured for grain crops, by marling too heavily.
On some of the land adjacent to the pieces measured in Exp.
10, and equally over-marled, very heavy red clover was
raised in 1830, by adding gypsum.

5th. All kinds of marl (or fossil shells) have sometimes
been injurious — but such effects have been more generally
experienced from the dry yellow marl, than from the blue
and wet. It is possible that some unknown ingredient in the
former may add to its hurtful power.

The inferences to be drawn from these facts are evident.
They direct us to avoid injury by applying marl lightly at
first, and to be still more cautious according to the existence
of the circumstances stated as increasing the tendency of marl
to do harm. Next, if the over-dose has already been given, to
forbid grazing entirely, and to furnish putrescent manure as

far as possible — or to omit one or two grain crops, so as to allow more vegetable matter to be fixed in the land — and to sow clover as soon as circumstances permit. One or more of these remedies have been used on most of my too heavily marled land, and with considerable, though not entire success. Putrescent manure, to the extent that it can be applied, is effectual. Other persons, who permitted close grazing, and adopted a more scourging rotation of crops, have suffered more, from lighter dressings of marl than mine.

But though the unlooked for damage sustained from this cause produced much loss and disappointment, and has greatly retarded the progress of my improvements, it did not stop my marling, nor abate my estimate of the value of the manure. If a cover of five hundred or six hundred bushels was so strong as to injure land of certain qualities, it seemed to be a fair deduction, that the benefit expected from so heavy a dressing, might have been obtained from half the quantity — if not on the first crop, at least on every one afterwards. *That* surely is nothing to be lamented. It also afforded some consolation, for the too heavy marlings already applied, that the soil was thereby fitted to seize and retain a greater quantity of vegetable matter, and ultimately, would reach a higher degree of fertility.

The cause of this disease is less apparent than its remedies. It is certain that it is not produced merely by the quantity of calcareous earth in the soil. If it were so, similar effects would always be found on soils containing far greater proportions of that earth. Such effects are not known to any extent, except on soils formerly acid, and made calcareous artificially. The small spots of land that nature has made excessively calcareous and sandy (as the specimen 4, page 43,) produce a pale feeble growth of corn, such as might be expected from a poor gravel — but whether the plants yield grain, or are barren, they show none of the peculiar symptoms of this disease which have been described.

By calculation, it appears that the heaviest dressing causing injurious consequences, mixed to the depth of five inches, has not given to the soil a proportion of calcareous earth equal to two per cent. This proportion is greatly exceeded in our best shelly land, and no such disease is found there, even when the rich mould is nearly all washed away, and the shells mostly left. Very fertile soils in France and England sometimes contain twenty or thirty per cent. of calcareous earth. Among the soils of remarkable good qualities analyzed by Davy, one is stated to contain about $28/100$, and another, which was eight-ninths of silicious sand, contained nearly $10/100$ of calcareous earth. Nor does he intimate that such proportions are very rare. Similar results have been stated, from analyses reported by Kirwan, Young, Bergman, and Rozier, (page 36,) and from all, the same deduction is inevitable, that much larger natural proportions of calcareous earth, than our diseased lands have received, are very common in France and England, without any such effect being produced.

From the numerous facts of which these are examples, it is certain that calcareous earth acting alone, or directly, has not caused this injury: and it seems most probable that the cause is some new combination of lime formed in acid soils only — and that this new combination is hurtful to grain under certain circumstances which we may avoid — and is highly beneficial to every kind of clover. Perhaps it is the *salt of lime*, formed by the calcareous manure with the acid of the soil, which not meeting with enough vegetable matter to combine with and fix in the soil, causes by its excess, all these injurious effects.

CHAPTER XIV

RECAPITULATION OF THE EFFECTS OF CAL-CAREOUS MANURES, AND DIRECTIONS FOR THEIR MOST PROFITABLE APPLICATION

PROPOSITION 5. *Continued.*

FROM the foregoing experiments may be gathered most of the effects, both injurious and beneficial, to be expected from calcareous manures, on the several kinds of soils there described. Information obtained from statements in detail of agricultural experiments, is far more satisfactory to an inquirer, than a mere report of the general opinions of the experimenter, derived from the results. But however valuable may be this mode of reporting facts, it is necessarily deficient in method, clearness, and conciseness. It may therefore be useful to bring together the general results of these experiments in a somewhat digested form, to serve as rules for practice. Some other effects will also be stated, which are equally established by experience, but which did not belong to any accurately observed experiment.

The results that have been reported confirm in almost every particular the chemical powers before attributed to calcareous manures, in the theory of their action. It is admitted that causes and effects were not always proportioned — and that sometimes trivial apparent contradictions were presented. But this is inevitable, even with regard to the best established doctrines, and the most perfect processes in agriculture. There are many practices universally admitted to be beneficial — yet there are none, which are not found sometimes useless, or hurtful, on account of some other attendant circumstance, which was not expected, and perhaps not discovered. Every application of calcareous earth to soil, is a chemical operation on a great scale. Decompositions and new combinations are produced, and in a manner generally conforming to the operator's expectations. But other and unknown agents may sometimes have a share in the process, and thus cause unlooked for results. Such differences between practice and theory have sometimes occurred in my use of calcareous manures (as may be observed in some of the reported experiments) but they have neither been frequent, uniform, nor important.

The benefit derived from marling will be in proportion to the vegetable or other putrescent matter given to the soil. It is essential that the cultivation should be mild, and no grazing permitted on poor lands. Wherever farm-yard manure is used, the land should be marled heavily, and if done previously, so much the better. The one manure cannot act by fixing the other, except so far as they are in contact, and both well mixed with the soil.

On *galled* spots, from which all the soil has been washed, and where no plant can live, the application of marl alone is utterly useless. Putrescent manures alone would there have but little effect, unless in great quantity, and would soon be all lost. But marl and putrescent matter together serve to form a new soil, and thus both are brought into useful action: the marl is made active, and the putrescent manure permanent. But though a fertile soil may thus be created, and fixed durably on *galls* otherwise irreclaimable, the cost will generally exceed the value of the land recovered, from the great quantity of putrescent matter required. Much of our acid hilly land, has been deprived by washing of a considerable portion of its natural soil, though not yet made entirely barren. The foregoing remarks equally apply to this kind of land, to the extent that its soil has been carried off. It will be profitable to apply marl to such land — but its effect will be diminished, in proportion to the previous removal of the soil. Calcareous soils are much less apt to wash, than other kinds, from the difference of texture. When a field that has been injured by washing, is marled, within a few years after, many of the old gullies will begin to produce vegetation, and show a soil gradually forming from the dead vegetables brought there by wind and rains, although no means should be used to aid this operation.

The effect of marling will be much lessened by the soil being kept under exhausting cultivation. Such were the circumstances under which we may suppose that marl was tried and abandoned many years ago, in the cases referred to in page 81. Supposing that marl was to enrich by direct action, it is most probable that it was applied to some of the poorest and most exhausted land, for the purpose of giving the manure a fair trial. The disappointment of such ill-founded expectations, was a sufficient reason for the experiment not being repeated, or being scarcely ever referred to again, except as evidence of the worthlessness of marl. Yet with proper views of the action of this manure, this experiment might have as well proved at first, the early efficacy and value of marl, as it now does its durability.

When acid soils are equally poor, the increase of the first crop from marling will be greater on sandy, than on clay soils; though the latter, by heavier dressings and longer time, may ultimately become the best land. The more acid the growth of any soil is, or would be, if suffered to stand, the more increase of crop may be expected from marl; which is directly the reverse of the effects of putrescent manures. The increase of the first crop on worn acid soil, I have never known under fifty per cent., and often is as much as one hundred — and the improvement continues to increase slowly under mild tillage. In this, and other general statements of effects, I suppose the land to bear not more than two crops in four years, and not to be subjected to grazing — and that a sufficient cover of marl has been laid on for use, and not enough to cause disease. It is true, that it is difficult, if not impossible, to fix that proper medium, varying as it may on every change of soil, of situation, and of the kind of marl. But whatever error may be made in the proportion of marl applied, let it be on the side of light dressing, (except where putrescent manures are also laid on) — and if less increase of crop is gained to the acre, the cost and labour of marling will be lessened in a greater proportion. If, after tillage has served to mix the marl well with the soil, sorrel should still show to any extent, it will sufficiently indicate that not enough marl had been applied, and that it may be added to, safely and profitably. If the nature of the soil, its condition and treatment, and the strength of the marl, all were known, it would be easy to direct the amount of a suitable dressing: but without knowing these circumstances, it will be safest to give two hundred and fifty or three hundred bushels to the acre of worn acid soils, and at least twice as much to newly cleared, or well manured land. Besides avoiding danger, it is more profitable to marl lightly at first on weak lands. If a farmer can carry out only ten thousand bushels of marl in a year, he will derive more product, and confer a greater amount of improvement, by spreading it over forty acres of the land intended for his next crop, than on twenty: though the increase to the acre, would probably be greatest in the latter case. By the lighter dressing, the whole farm will be marled, and be storing up vegetable matter, in half the time that it could be marled at double the rate.

The greater part of the calcareous earth applied at one time cannot begin to act as manure, before several years have passed, owing to the coarse state of many of the shells, and the want of thoroughly mixing them with the soil. Therefore, if enough marl is applied to obtain its full effect on the first course of crops, there will be too much afterwards.

Perhaps the greatest profit to be derived from marling, though not the most apparent, is on such soils as are full of

wasting vegetable matter. Here the effect is mostly preserv-
ative, and the benefit may be great, even though the increase
of crop may be very inconsiderable. Putrescent manure laid
on any acid soil, or the natural vegetable cover of those newly
cleared, without marl, would soon be lost, and the crops
reduced to one half, or less. But when marl is previously
applied, this waste of fertility is prevented; and the estimate
of benefit should not only include the actual increase of crop
caused by marling, but as much more as the amount of the
diminution, which would otherwise have followed. Every
intended clearing of woodland, and especially of that under
a second growth, ought to be marled before cutting down —
and it will be still better, if it can be done several years before.
If the application is delayed until the new land is brought
under cultivation, though much putrescent matter will be
saved, still more must be wasted. By using marl some years
before obtaining a crop from it, as many more growths of
leaves will be converted to useful manure, and fixed in the
soil — and the increased fertility will more than compensate
for the delay. By such an operation, we make a loan to the
soil, with a distant time for payment, but an ample security,
and at a high rate of compound interest.

Some experienced cultivators have believed that the most
profitable way to manage pine old fields, when cleared of
their second growth, was to cultivate them every year, until
worn out — because, as they said, such land would not last
much longer, no matter how mildly treated. This opinion,
which seems so absurd — and is opposed to all the received
rules for good husbandry, is considerably supported by the
properties, which are here ascribed to such soils. When these
lands are first cut down, an immense quantity of vegetable
matter is accumulated on the surface — which, notwithstanding its accompanying acid quality, is capable of making
two or three crops nearly or quite as good as the land was
ever able to bear. But as the soil has no power to retain this

vegetable matter, it will begin rapidly to decompose and waste, as soon as exposed to the sun, and will be lost, except so much as is caught while escaping, by the roots of growing crops. The previous application of marl, would make it profitable in these, as well as other cases, to adopt a mild and meliorating course of tillage.

Less improvement will be obtained by marling worn soils of the kind called "free light land," than other acid soils which originally produced much more sparingly. The early productiveness of this kind of soil, and its rapid exhaustion by cultivation, at first view seem to contradict the opinion that durability and the ease of improving by putrescent manures, are proportioned to the natural fertility of the soil. But a full consideration of circumstances will show that no such contradiction exists.

In defining the term *natural fertility*, it was stated that it should not be measured by the earliest products of a new soil, which might be either much reduced, or increased, by temporary causes. The early fertility of free light land is so rapidly destroyed, as to take away all ground for considering it as fixed in, and belonging to the soil. It is like the effect of dung on the same land afterwards, which throws out all its effect in the course of one or two years, and leaves the land as poor as before. But still it needs explanation why so much productiveness can at first be exerted by any acid soil, as in those described in the 14th Experiment. The cause may be found in the following reasons. These soils, and also their subsoils, are principally composed of coarse sand, which makes them of more open texture than best suits pine, and (when rich enough) more favourable to other trees, the leaves of which have no natural acid, and therefore decompose more readily. As fast as the leaves rot, they are of course exposed to waste — but the rains convey much of their finer parts down into the open soil, where the less degree of heat retards their final decomposition. Still this enriching matter is liable to be further decomposed, and to final waste: but though continually wasting, it is also continually added to by the rotting leaves above. The shelter of the upper coat of unrotted leaves, and the shade of the trees, cause the first as well as the last stages of decomposition to proceed slowly, and to favour the mechanical process of the products being mixed with the soil. But there is no chemical union of the vegetable matter with the soil. When the land is cleared, and opened by the plough, the decomposition of all the accumulated vegetable matter is hastened by the increased action of sun and air, and in a short time converts every thing into food for plants. This abundant supply suffices to produce two or three fine crops. But now, the most fruitful source of vegetable matter has been cut off — and the soil is kept so heated (by its open texture) as to be unable to hold enriching matters, even if they were furnished. The land soon becomes poor, and must remain so, as long as these causes operate, even though cultivated under the mildest rotation. When the transient fertility of such a soil is gone, its acid qualities (which were before concealed in some measure by so much enriching matter,) become evident. Sorrel and broom grass cover the land — and if allowed to stand, pines will take complete possession, because the poverty of the soil leaves them no rival to contend with.

Marling deepens cultivated sandy soils, even lower than the plough may have penetrated. This was an unexpected result, and when first observed, seemed scarcely credible. But this effect also is a consequence of the power of calcareous earth of fixing manures. As stated in the foregoing paragraph, the soluble and finely divided particles of rotted vegetable matters are carried by the rains below the soil: but as there is no calcareous earth there to fix them, they must again rise in a gaseous form, after their last decomposition, unless previously taken up by growing plants. But after the soil is marled, calcareous as well as putrescent matter is carried down by the rains as far as the soil is open enough for them to pass. This will always be as deep as the ploughing has been, and in loose earth, somewhat deeper — and the chemical union formed between these different substances, serves to fix both, and thus increases the depth of the soil. This effect is very different from the deepening of a soil by letting the plough run into the barren subsoil. If by this mechanical process, a soil of only three inches is increased to five, as much as it gains in depth, it loses in richness. But when a marled soil is deepened gradually, its dark colour and apparent richness is increased, as well as its depth. Formerly single-horse ploughs were used to break all my acid soils, and even they would often turn up subsoil. The average depth of soil on old land did not exceed three inches, nor two on the newly cleared. Even before marling was commenced, my ploughing had generally sunk into the subsoil — and since 1825, most of this originally thin soil has required three mules, or two good horses to a plough, to break the necessary depth. The soil is now from five to seven inches deep generally, from the joint operation of marling and deepening the ploughing a little in the beginning of every course of crops.

On acid soils without manure, it is scarcely possible to raise red clover — and even with every aid from putrescent manure, the crop will be both uncertain and unprofitable. The recommendation of this grass as part of a general system of improvement, by the author of *Arator*,[1] is sufficient to prove that his improvements were made on soils far better than such as are general. After much waste of seed and labour, and years of disappointed efforts, I gave up clover as utterly hopeless. After marling the fields on which raising clover had been vainly attempted, there arose from its scattered and feeble remains, a growth which served to prove that its cultivation would then be safe and profitable. It has since been gradually extended over most of the fields. It will stand well, and maintain a

[1][John Taylor of Caroline County, Virginia.]

healthy growth on the poorest marled land: but the crop is too
scanty for mowing, or perhaps for profit of any kind on most
light soils, unless aided by gypsum. Newly cleared lands yield
better clover than the old, though the latter may produce as
heavy grain crops. The remarkable crops of clover raised
on very poor clay soils, after marling, have been already
described. This grass, even without gypsum, and still more if
aided by that manure, may add greatly to the improving power
of marl: but it will do more harm than service, if we greedily
take from the soil too large a share of this supply of putrescent
matter.

Some other plants less welcome than clover, are still more
favoured by marling. Greensward, blue grass, wire grass,
and partridge pea, will soon extend so as to be not less impedi-
ments to tillage, than evidences of an entire change in the
character and power of the soil.

With all the increase of products that I have ascribed to
marling, the heaviest crops stated may appear inconsiderable
to farmers who till soils more favoured by nature. Corn
yielding twenty-five or thirty bushels to the acre, is doubled
by many natural soils in the western states, and ten or twelve
bushels of wheat, will still less compare with the product of
the best limestone clay land. The cultivators of our poor
region, however, know that such products, without any future
increase, would be a prodigious addition to their present
gains. Still it is doubtful whether these rewards are sufficiently
high to tempt many of my countrymen speedily to accept
them. The opinions of many farmers have been so long fixed,
and their habits are so uniform and unvarying, that it is
difficult to excite them to adopt any new plan of improvement, except by promises of profits so great, that an uncommon share of credulity would be necessary to expect their
fulfilment. The nett profits of marling, if estimated at twenty
or even fifty per cent. per annum on the expense, forever —
or the assurance by good evidence, of doubling the crops of

a farm in twelve years — will scarcely attract the attention of those who would embrace without any scrutiny a plan that promised five times as much. Hall's scheme for cultivating corn was a stimulus exactly suited to their lethargic state: and that impudent impostor found many steady oldfashioned farmers willing to pay for his directions for making two thousand five hundred bushels of corn, with the hand labour of only two men.

CHAPTER XV

THE PERMANENCY OF CALCAREOUS MANURES

PROPOSITION 5. *Continued.*

IT has been stated that the ground on which an old experiment was made and abandoned as a failure more than fifty years ago, still continues to show the effects of marl. Lord Kames mentions a fact of the continued beneficial effect of an application of calcareous manure, which was known to be one hundred and twenty years old.[1] Every author who has treated of manures of this nature, attests their long duration: but when they say that they will last twenty years, or even one hundred and twenty years, it amounts to the admission that at some future time the effects of these manures will be lost. This I deny — and from the nature and action of calcareous earth, claim for its effects a duration that will have no end.

If calcareous earth applied as manure is not afterwards combined with some acid in the soil, it must retain its first form, which is as indestructible, and as little liable to be wasted by any cause whatever, as the sand and clay that form the other earthy ingredients of the soil. The only possible vent for its loss, is the very small proportion taken up by the roots of plants, which is so inconsiderable as scarcely to deserve naming.

Clay is a manure for sandy soils, serving to close their too open texture. When so applied, no one can doubt but that this effect of the clay will last as long as its presence. Neither can calcareous earth cease to exert its peculiar powers as a manure, any more than clay can, by the lapse of time, lose

[1] [Henry Home (Lord Kames)] *The Gentleman Farmer* [2nd ed., Edinburgh, 1779], p. 266.

its power of making sands more firm and adhesive. Making due allowance for the minute quantity drawn up into growing plants, it is as absurd to assert that the calcareous earth in a soil, whether furnished by nature or not, can be exhausted, as that cultivation can deprive a soil of its sand or clay.

But on my supposition that calcareous earth will change its form by combining with acid in the soil, it may perhaps be doubted whether it is equally safe from waste under its new form. It must be admitted, that the permanency of this compound cannot be proved by its insolubility, or other properties, because neither the kind nor the nature of the salt itself is yet known. But judging from the force with which good neutral soils resist the exhaustion of their fertility, and their always preserving their peculiar character, it cannot be believed that the calcareous earth once present, was lessened in durability by its chemical change of form. It has been contended that the action of calcareous earth is absolutely necessary to make a poor acid soil fertile: but it does not thence follow that other substances, and particularly this salt of lime, may not serve as well to preserve the fertility bestowed by calcareous earth. All that is required for this purpose, is the power of combining with putrescent matter, and thereby fixing it in the soil: and judging solely from effects, this power seems to be possessed in an eminent degree by this new combination of lime. If this salt is the oxalate of lime, (as there is most reason to believe,) it is insoluble in water, and consequently safe from waste — and the same property belongs to most other combinations of lime with vegetable acid. The acetate of lime is soluble in water, and while alone, might be carried off by rains. But if it combines with putrescent matter, by chemical affinity, its previous solubility will no longer remain. Copperas is easily soluble: but when it forms one of the component parts of ink, it can no longer be separately dissolved by water, or taken away from the colouring matter combined with it. In rich limestone soils, and some of our best river lands, in which no calcareous earth remains, we may suppose that its change of form took place centuries ago. Yet however scourged and injured by cultivation, they still show as strongly as ever those qualities which were derived from their former calcareous ingredient. When the dark colour of such soils, their power of absorption, and of holding manures, their friability, and their peculiar fitness for clover and certain other plants, are no longer to be distinguished, then, and not before, may the salt of lime be considered as lost to the soil.

If we keep in mind the mode by which calcareous manure acts, its effects may be anticipated for a much longer time than my experience extends. Let us trace the supposed effects, from the causes, on an acid soil, kept under meliorating culture. As soon as applied, the calcareous earth combines with all the acid then present, and to that extent, is changed to the vegetable salt of lime. The remaining calcareous earth continues to take up the after formations of acid, and (together with the salt so produced,) to fix putrescent manures, as fast as these substances are presented, until all the lime has been combined with acid, and all their product combined with putrescent matter. Both those actions then cease. During all the time necessary for those changes, the soil has been regularly increasing in productiveness; and it may be supposed that before their completion, the product had risen from ten to thirty bushels of corn to the acre. The soil has then become neutral. It can never lose its ability (under the mild rotation supposed,) of producing thirty bushels — but it has no power to rise above that product. Vegetable food continues to form, but is mostly wasted, because the salt of lime is already combined with as much as it can act on; and whatever excess of vegetable matter remains on the soil, is kept useless by acid also newly formed, and left free and noxious, as before the application of calcareous earth. But though this excess of acid may balance and keep useless the excess of vegetable matter, it cannot affect the previously fixed fertility, nor lessen the power of the soil to yield its then maximum product of thirty bushels. In this state of things, sorrel may again begin to grow, and its return may be taken as notice that a new marling is needed, and will afford additional profit, in the same manner as before, by destroying the last formed acid, and fixing the last supply of vegetable matter. Thus perhaps five or ten bushels more may be added to the previous product, and a power given to the soil gradually to increase as much more, before it will stop again, for similar reasons, at a second maximum product of forty or fifty bushels. I pretend not to fix the time necessary for the completion of one or more of these gradual changes: but as the termination of each, and the consequent additional marling, will add new profits, it ought to be desired by the farmer, instead of his wishing that his first labour of marling each acre, may also be the last required. Every permanent addition of five bushels of corn to the previous average crop, will more than repay the heaviest expenses that have yet been encountered in marling. But whether a second application of marl is made or not, I cannot imagine such a consequence as the actual decrease of the product once obtained. My earliest marled land has been severely cropped, compared to the rotation supposed above, and yet has continued to improve, though at a slow rate. The part first marled in 1818, has since had only four years of rest in fifteen; and has yielded nine crops of grain, one of cotton, and one year clover, twice mowed. This piece, however, besides being sown with gypsum, (with little benefit,) once received a light cover of rotted corn-stalk manure. The balance of the same piece of land (Exp. I.) was marled for the crop of 1821 — has borne the same treatment since, and has had no other manure, except gypsum once, (in 1830,) which acted well. These periods of twelve and fifteen years are very short to serve as grounds to decide on the eternal duration of a manure. But it can scarcely be believed that the effect of any temporary manure, would not have been somewhat abated by such a course of severe tillage. Under milder treatment, there can be no doubt but there would have been much greater improvement.

If subjected to a long course of the most severe cultivation, a soil could not be deprived of its calcareous ingredient whether natural or artificial: but though still calcareous, it would be in the end, reduced to barrenness, by the exhaustion of its vegetable matter. Under the usual system of exhausting cultivation, marl certainly improves the product of acid soils, and may continue to add to the previous amount of crop, for a considerable time: yet the theory of its action instructs us, that the ultimate result of marling under such circumstances, must be the more complete destruction of the land, by enabling it to yield all its vegetable food to growing plants, which would have been prevented by the continuance of its former acid state. An acid soil yielding only five bushels of corn, may contain enough food for plants to bring fifteen bushels — and its production will be raised to that mark, as soon as marling sets free its dormant powers. But a calcareous soil reduced to a product of five bushels, can furnish food for no more, and nothing but an expensive application of putrescent manures, can render it worth the labour of cultivation. Thus it is, that soils, the improvement of which is most hopeless without calcareous manures, will be the most certainly improved with profit by their use. [Appendix. G.]

CHAPTER XVI

THE EXPENSE AND PROFIT OF MARLING

PROPOSITION 5. *Concluded.*

AT this time there are but few persons among us who doubt the great benefit to be derived from the use of marl: and many of those who ten years ago deemed the practice the result of folly, and a fit subject for ridicule, now give that manure credit for virtues which it certainly does not possess, and from their manner of applying it, seem to believe it a universal cure for sterility. Such erroneous views have been a principal cause of the many injudicious and even injurious applications of marl. It is as necessary to moderate the ill-founded expectations which many entertain, as to excite the too feeble hopes of others.

The improvement caused by marling, and its permanency, have been established beyond question. Still the improvement may be paid for too dearly — and the propriety of the practice must depend entirely on the amount of its clear profits, ascertained by fair estimates of the expenses incurred.

With those who attempt any calculations of this kind, it is very common to set out on the mistaken ground that the expense of marling should bear some proportion to the selling price of the land: and without in the least underrating the effects of marl, they conclude that the improvement cannot justify an expense of six dollars on an acre of land that would not previously sell for four dollars. Such a conclusion would be correct if the land was held as an article for sale, and intended to be disposed of as soon as possible: as the expense in that case might not be returned in immediate profit, and certainly would not be added to the price of the land by the purchaser, under present circumstances. But if the land is held as a possession of any permanency, its previous price, or its subsequent valuation, has no bearing whatever on the amount which it may be profitable to expend for its improvement. Land that sells at four dollars, is often too dear at as many cents, because its product will not pay the expense of cultivation. But if by laying out for the improvement ten dollars, or even one hundred dollars to the acre, the average increased annual profit would certainly and permanently be worth ten per cent. on that cost of improvement, then the expenditure would be highly expedient and profitable. We are so generally influenced by a rage for extending our domain, that another farm is often bought, stocked and cultivated, when a liberal estimate of its expected products, would not show an annual clear profit of three per cent.: and any one would mortgage his estate to buy another thousand acres, that was supposed fully capable of yielding ten per cent. on its price. Yet the advantage would be precisely the same, if the principal money was used to enrich the land already in possession, (without regard to its extent, or previous value,) with equal assurance of its yielding the same amount of profit.

Nothing is more general, or has had a worse influence on the state of agriculture, than the desire to extend our cultivation, and landed possessions. One of the consequences of this disposition, has been to give an artificial value to the poorest land, considered as merely so much territory, while various causes have concurred to depress the price of all good soils much below their real worth. Whatever a farm will sell for, fixes its value as merchandise; but by no means is it a fair measure of its value as permanent farming capital.

The true value of land, and also of any permanent improvements to land, I would estimate in the following manner. Ascertain as nearly as possible the average clear and permanent income, and the land is worth as much money as would securely yield that amount of income, in the form of interest.

For example, if a field brings ten dollars average value of crops to the acre, in every course of a four shift rotation, and the average expense of every kind necessary to carry on the cultivation, is also ten dollars — then the land yields nothing, and is worth nothing. If the average clear profit was two dollars and forty cents in the term, or only sixty cents a year, it would raise the value of the land to ten dollars — and if six dollars could be made annually, clear of all expense, it is equally certain that one hundred dollars would be the fair value of the acre. Yet if lands of precisely these rates of profit were offered for sale at this time, the poorest would probably sell for two dollars, and the richest for less than thirty dollars. In like manner, if any field that paid the expense of cultivation before, has its average annual nett product increased six dollars for each acre, by some permanent improvement, the value thereby added to the field is one hundred dollars the acre, without regard to its former worth. Let the cost and value of marling be compared by this rule, and it will be found that the capital laid out in that mode of improvement will seldom return an annual interest of less than twenty per cent. — that it will more often equal forty — and sometimes reach even one hundred per cent. of annual and permanent interest on the investment. The application of this rule for the valuation of such improvements, will raise them to such an amount, that the magnitude of the sum may be deemed a sufficient contradiction of my estimates. But before this mode of estimating values is rejected, merely on the supposed absurdity of an acid soil being considered as raised from one dollar to thirty dollars per acre, by a single marling, let it at least be examined, and its fallacy exposed.

I admit the practical difficulty of applying this rule, however certain may be its theoretical truth. It is not possible to fix on the precise clear profit of any farm to its owner and cultivator; and any error made in these premises, is increased sixteen fold in the amount of value founded on them.

Still we may approximate the truth with most certainty by using this guide. The early increase of crop from marling, will in most cases be an equal increase of clear profit, (for the subsequent improvement and the additional offal will pay for the increase of labour,) — and it is not very difficult to fix a value for that actual increase of crop, and thereby to estimate the capital value of the improvement.

This mode of valuing land, under a different form, is universally received as correct in England. Cultivation there is carried on almost entirely by tenants: and the annual rent which any farm brings on a long lease, fixes beyond question what is its annual clear profit to the owner. The price, or value of land, is generally estimated at so many "years' purchase," which means as many years' rent as will return the purchaser's money. There, the interest of money being low, increases the value of land according to this mode of estimation; and it is generally sold as high as twenty years' purchase. My estimate is less favourable for raising the value of our lands, as it fixes them at sixteen and a half years' purchase.

But an objector may ask, "if any poor land is raised in value (according to this estimate) from one dollar to thirty by marling, would a purchaser make a judicious investment of his capital, by buying this improved land at thirty dollars?" — I would answer in the affirmative, if our view was confined to this particular means of investing farming capital. The purchaser would get a clear interest of six per cent. — which is always a good return from land, and is twice as much as all Lower Virginia now yields. But if such a purchase is compared with other means of acquiring land so improved, it would be extremely injudicious — because thirty dollars expended in purchasing and marling such land, would serve both to acquire and improve five or six acres.

Estimates of the expenses required for marling are commonly erected on as improper grounds, as those of its profits. We never calculate the cost of any old practice. We are content to clear woodland that afterwards will not pay for the expense of tillage — to keep under the plough, land reduced to five bushels of corn to the acre — to build and continue to repair miles of useless and perishable fences — to make farm-yard manure (though not much of this fault,) and apply it to acid soils — without once calculating whether we lose or gain by any of these operations. But let any new practice be proposed, and every one begins to count its cost — and on such erroneous premises, that if applied to every kind of farm labour, the estimate would prove that the most fertile land known, could scarcely defray its expenses.

According to estimates made with much care and accuracy, the cost of an uncommonly expensive job of marling, four thousand and thirty-six bushels in quantity, in 1824, amounted to five dollars and thirty-five cents the acre, for five hundred and ninety-eight bushels of marl. This quantity was much too great: four hundred bushels would have been quite enough for safety and profit, and would have reduced the whole expense, including every necessary preparation, to three dollars and fifty-eight cents the acre. The earth which was taken off, to uncover the bed of marl, was considerably thicker than the marl itself. The road from the pit ascended hills amounting to forty feet of perpendicular elevation — and the average distance to the field was eight hundred and forty-seven yards.

In 1828, I began to marl another tract of land, where the difficulties were less. The labour bestowed served to carry out and spread six thousand eight hundred and ninety-two tumbril loads, on one hundred and twenty acres of land, being an average of two hundred and fifty-nine bushels to the acre. The exhausted state of the soil made heavier dressing unsafe. The whole expense of the operation, including all the preparatory labour, amounted to two dollars and eight cents for each acre marled — or $83\frac{3}{100}$ of a cent for each heaped bushel of marl. [Appendix. H.]

It is impossible to carry on marling to advantage, or with

any thing like economy, unless it is made a regular business, to be continued throughout the year, or a specified portion of it, by a labouring force devoted to that purpose, and not allowed to be withdrawn for any other. Instead of proceeding on this plan, most persons who have begun to marl, attempted it in the short intervals of leisure, afforded between their different farming operations — and without lessening for this purpose, the extent of their usual cultivation. Let us suppose that preparations have been made, and on the first opportunity, a farmer commences marling with zeal and spirit. But every new labour is attended by causes of difficulty and delay, and a full share of these will be found in the first few days of marling. The road is soft for want of previous use, and if the least wet, soon becomes miry. The horses, unaccustomed to carting, balk at the hills, or only carry half loads. Other difficulties occur from the awkwardness of the labourers, and the inexperience of their master — and still more from the usual unwillingness of his overseer to devote any labour to improvements which are not expected to add to the crop of that year. Before matters can get straight, the leisure time is at an end: the work is stopped, and the road and pit are left to get out of order, before making another attempt some six months after, when all the same vexatious difficulties are again to be encountered.

If only a single horse was employed in drawing marl throughout the year, at the moderate allowance of two hundred working days, and one hundred bushels carried out for each, his year's work would amount to twenty thousand bushels, or enough for more than sixty acres. This alone would be a great object effected. But besides, this plan would allow the profitable employment of additional labour. When at any time, other teams and labourers could be spared to assist, though for only a few days, every thing is ready for them to go immediately to work. The pit is drained, the road is firm, and the field marked off for the loads. In this way, much labour may be obtained in the course of the year, from teams that would otherwise be idle, and labourers whose other employments would be of but little importance. The spreading of marl on the field, is a job that will always be ready to employ any spare labour: and throwing off the covering earth from an intended digging of marl, may be done, when rain, snow, or severe cold, have rendered the earth unfit for almost every other kind of labour.

Another interesting question respecting the expense of this improvement, is, what distance from the pit may marl be profitably carried? If the amount of labour necessary to carry it half a mile is known, it is easy to calculate how much more will be required for two or three miles. The cost of teams and drivers is in proportion to the distance travelled — but the pit and field labour, is not affected by that circumstance. At present, when so much poor land, abundantly supplied with fossil shells, may be bought at from two dollars to four dollars the acre, a farmer had better buy and marl a new farm, than to move marl even two miles to his land in possession. But this would be merely declining one considerable profit, for the purpose of taking another much greater. Whenever the value of marl is properly understood, and our lands are priced according to their improvements, or their capability of being improved from that source, as must be the case hereafter, then this choice of advantages will no longer be offered. Then rich marl will be profitably carted miles from the pits, and perhaps conveyed by water as far as it may be needed. A bushel of such marl as the bed on James River, described page 112, is as rich in calcareous earth alone as a bushel of slacked lime will be after it becomes carbonated — and the greater weight of the first, is a less disadvantage for water carriage, than the price of the latter. Farmers on James River who have used lime as manure to great extent and advantage, might more cheaply have moved rich marl twenty miles by water, as it would cost nothing but the labour required to dig and move it. Either lime or good marl may hereafter be profitably distributed over a remote strip of poor land, by means of the rail road now constructing from Petersburg to the Roanoke: provided the proprietors do not imitate the over greedy policy of the legislature of Virginia, in imposing tolls on manures passing through the James River canal. If there was no object whatever in view, but to draw the greatest possible income from tolls on canals and roads, true policy would direct that all manures should pass from town to country toll free. Every bushel of lime, marl, or gypsum, thus conveyed, would be the means of bringing back in future time, more than as much wheat or corn — and there would be an actual gain in tolls, besides the twenty fold greater increase to the wealth of individuals and the state.[1] Wood ashes, after being deprived of their potash, have calcareous earth, and a smaller proportion of phosphate of lime, as their only fertilizing ingredients; and both together do not commonly make more, than there is of calcareous earth in the same bulk of good marl. Yet drawn ashes have been purchased largely from our soap factories, at four cents the bushel, and carried by sea to be sold for manure to the farmers of Long Island. Except for the proportion of phosphate of lime which they contain, drawn ashes are simply artificial marl — more fit for immediate action,

[1] Since writing the above passage, several new schemes for rail roads in Lower Virginia have been brought forward, and strongly advocated. Among them are proposed rail roads from Richmond to Yorktown, and from Richmond to Westover. It is not my intention to offer an opinion on the advantages or disadvantages of any of these plans. But should either be constructed, I will venture to prophecy that an immense amount of transportation of rich marl, and lime, will take place, provided that the expense shall not exceed that on canals navigated toll free. The line for either of these routes must necessarily be located on poor ridge land, now perfectly worthless, but of the kind most susceptible of being highly improved by marling. The lower end of both these lines, but particularly that to Yorktown, is convenient to marl of the best quality. Whenever from twenty to forty square miles of this barren land shall produce thirty bushels of corn to the acre, it will cause a great increase of profit to the road, and form no contemptible addition to the trade of Richmond.

by being finely divided, but weaker than our best beds of fossil shells.

The argument in support of the several propositions which have been discussed through so many chapters, is now concluded. However unskilfully, I flatter myself that it has been effectually used; and that the general deficiency in our soils of calcareous earths — the necessity of supplying it — the profit by that means to be derived — and the high importance of all these considerations — have been established too firmly to be shaken by either arguments or facts.

CHAPTER XVII

DIRECTIONS FOR DIGGING AND CARTING MARL

THE great deposit of fossil shells, which custom has mis-called marl, is in many places exposed to view, in most of the lands that border on our tide-waters, and on many of their small tributary streams. Formerly, it was supposed to be limited to such situations: but since its value as a manure has caused it to be more noticed, and sought after, marl has been found in many other places. It is often discovered by the digging of wells, but lying so deep, that its value must be more highly estimated than at present, before it will be dug for manure. From all the scattered evidences of the presence of this deposit, it may be inferred, that it lies beneath nearly every part of our country between the sea and the granite ridge which forms the falls of all our rivers. It is exposed, where it rises, and where cut through by the deep ravines of hilly land, and the courses of rivers — and concealed by its dips, and the usual level surface of the country. The rich tracts of neutral soil on James River, such as Shirly, West-over, Brandon, and Sandy Point, seem to have been formed by alluvion, which may be termed recent, compared to that of our district in general: and in these, no marl has been found, though it is generally abundant in the adjacent higher lands. Fresh-water muscle shells are sometimes found in thin layers (from a few inches to two feet thick) both on those lands, and others — but generally near the surface, and always far above the deposit of sea shells, found under the high land. These two layers of different kinds of shells are separated by a thickness of many feet of earth, containing no shells of any kind. From these appearances, it would seem that this tract of country was, for ages, the bottom of the sea — then covered by earth — then the bottom of a fresh-water lake — and finally made dry land. Muscle shells are richer than the others, as they contain much gelatinous and enriching animal matter. On this account, the earth with which muscle shells are found mixed, is a rich black mould — while that mixed with fossil sea shells, is quite barren. Most persons consider these beds of muscle shells as artificially formed by the Indians, who are supposed to have collected the muscles, for food, and left the shells, where the fish were consumed. But besides several other reasons for opposing this opinion, it is sufficient to examine the shells as they now lie in layers in the earth, to be satisfied, that they must have been in water, when deposited in their present situation.

More than forty kinds of sea shells are found in the beds of marl that I have worked. Generally the shells are whole, but are much broken by digging, and the after operations. The white shells are rapidly reduced, after being mixed with an acid soil — but some grey kinds, as scallop, and a variety of oyster, are so hard as to be very long before they can act as manure. Some beds, and they are generally the richest, have scarcely any whole shells, but are formed principally of small broken fragments. Of course the value of marl as a manure depends in some measure on what kinds of shells are most numerous, and their state of division, as well as upon the total amount of the calcareous earth contained. The last is however by far the most important criterion of its value. The most experienced eye may be much deceived in the strength of marl, and still more gross and dangerous errors would be made by an inexperienced marler. The strength of a body of marl often changes materially in sinking a foot in depth — although the same changes may be expected to occur very regularly, in every pit sunk through the same bed. Whoever uses marl, ought to know how to analyze it, which a little care will enable any one to do with sufficient accuracy, by either of the two methods described in the 5th chapter. But the analysis of marl (that is, merely ascertaining its proportion of calcareous earth) by the pneumatic apparatus, is by far the most accurate, expeditious, and even cheapest method, if many and frequent trials are desired.

Our beds of marl are either of a blue, or a yellowish colour. The colour of the first seems to have some connexion with the presence of water, as this kind is always kept wet, by water slowly oozing through it. The yellow marl is sometimes wet, but more generally dry, and therefore easier to work. Unless very poor, all marls are sufficiently firm and solid for the sides of the pit to stand, when dug perpendicularly.

Where a bed of marl is dry and not covered by much earth, no directions are required for the pit work — except it be, that the pit should be long enough to allow the carts to descend to the bottom (when finished) and to rise out on a slope sufficiently gradual. This will prevent the necessity of twice handling the marl, by first throwing it out of the pit, and then into the carts, which must be done, if the pit is made too short, or its ends too steep, for the loads to be drawn out. No machine or contrivance will raise marl from the bottom of a pit, or a valley, so well as a horse-cart — and no pains will be lost, in enlarging the pit, and graduating the ascent out of it to attain that object.

As marl usually shows on a hill-side, but little earth has to be moved to uncover the first pit. But the next, and each successive cover of earth, will be more thick, until it may be necessary to abandon that place and begin again elsewhere. But the quantity of covering earth need not be regarded as a serious obstacle, if it is not thicker than the marl below it. While that is the case, one pit completed will receive all the earth thrown from an equal space, for commencing another. When this proportion of earth is exceeded, it is necessary to carry it farther, by either carts or scrapers, and the labour is greatly increased.

For any extensive operation, it is much cheaper to take off a cover of earth, twelve feet thick, to obtain marl of equal depth, than if both the covering earth and marl were only three feet each. Whether the cover be thick or thin, two parts of the operation are equally troublesome, viz. to take off the mat of roots, and perhaps some large trees on the surface soil, and to clean off the surface of the marl, which is sometimes very irregular. The greater part of the thickest cover would be much easier to work. But the most important advantage in taking off earth of ten or more feet in thickness, is saving digging, by causing the earth to come down by its own weight. If time can be allowed to aid this operation, the driest earth will mostly fall, by being repeatedly undermined a little. But this is greatly facilitated by the oozing water, which generally fills the earth lying immediately on beds of wet marl. In uncovering a bed of this description, where the marl was to be dug fourteen feet, and ten to twelve feet of earth to remove, my labour was made ten-fold heavier, by digging altogether. The surface bore living trees, and was full of roots — there was enough stone to keep the edges of the grubbing hoes battered — and small springs and oozing water came out every where, after digging a few feet deep. A considerable part of the earth was a tough, sticky clay, kept wet throughout, and which it was equally difficult to get on the shovels, and to get rid of. Some years after, another pit was uncovered on the same bed, and under like circumstances, except that the time was the last of summer, and there was less water oozing through the earth. This digging was begun at the lowest part of the earth, which was a layer of sand, kept quite wet by the water oozing through it. With gravel shovels, this was easily cut under from one to two feet along the whole length of the old pit — and as fast as was desirable, the upper earth, thus undermined, fell, into the old pit at first, and afterwards, when that did not take place of itself, the fallen earth was easily thrown there by shovels. As the earth fell separated into small but compact masses, it was not much affected by the water, even when it remained through the night, before being shoveled away. No digging was required, except this continued shoveling out the lowest sand stratum, and whether clay, or stones, or roots, were mixed with the falling earth, they were easy to throw off. The numerous roots which were so troublesome in the former operation, were now an advantage, as they supported the earth sufficiently to let it fall only gradually and safely; and before the roots fell, they were almost clear of earth. The whole body of earth, with all its difficulties, was moved off as easily as the driest could have been by digging altogether.

In working a pit of wet marl, no pains should be spared to drain it as effectually as possible. Very few beds are penetrated by veins of running water, which would deserve the name of springs — but water oozes very slowly through every part of wet marl, and bold springs often burst out immediately over its surface. After the form of the pit, and situation of the road are determined, a ditch to receive and draw off all the water, should be commenced down the valley, as low as the bottom of the pit is expected to be, and opened up to the work, deepening as it extends, so as to keep the bottom of the ditch on the same level with the bottom of the marl. It may be cheaper, and will serve as well, to deepen this ditch as the deepening of the pit proceeds. After the marl is uncovered the full size intended for the pit, (which ought to be large enough for carts to turn about on,) a little drain of four or five inches wide, and as many deep, or the size made by the grubbing hoe used to cut it, should be carried all around to intercept the surface or spring water and conduct it to the main drain. The marl will now be dry enough for the carts to be brought on and loaded. But as the digging proceeds, oozing water will collect slowly, and aided by the wheels of loaded carts, the surface of the firmest marl would soon be rendered a puddle, and then a quagmire. This may easily be prevented by the inclination of the surface. The first course dug off, should be much the deepest next the surface drain, (leaving a margin of a few inches of finer marl, as a bank to keep in the stream) so that the digging shall be the lowest around the outside, and gradually rise to the middle of the area. Whatever water may find its way within the work, whether from oozing, rain, or accidental burstings of the little surface drain, will run to the outside, the dip of which should lead to the lower main drain. After this form is given to the surface of the area, very little attention is required to preserve it — for if the successive courses are dug of equal depth from side to side, the previous dip will not be altered. The sides or walls of the pit should be cut something without the perpendicular, so that the pit is made one or two feet wider at bottom than top. The usual firm texture will prevent any danger from this overhanging shape, and several advantages will be gained from it. It gives more space for work — prevents the wheels running on the lowest and wettest parts — allows more earth to be disposed of, in opening for the next pit — and prevents that earth tumbling into the next digging, when the separating wall of marl is cut away. The upper drain of the pit, which takes the surface water, will hang over the one below, kept for the oozing water. The first remains unaltered throughout the job, and may still convey the stream, when six feet above the heads of the labourers in the pit. The lower drain of course sinks with the digging. Should the pit be dug deeper than the level of the receiving ditch can be sunk, a wall should be left between, and the remainder of the oozing water must be conducted to a little basin near the wall, and thence baled or pumped into the receiving ditch. The passage for the carts to ascend from the pit should be kept on a suitable slope — and the marl forming that slope may be cut out in small pits, after the balance has been completed.

If the marl is so situated that carts cannot be driven as

low as the bottom, then the area must be cut out in small pits, beginning at the back part, and extending as they proceed, towards the road leading out of the pit.

On high and broken land, marl is generally found at the bottom of ravines, and separated from the field where it is to be carried, by a high and steep hill-side. The difficulty of cutting roads in such situations, is much less than any inexperienced person would suppose. We cannot get rid of any of the actual elevation — but the ascent may be made as gradual as is desired, by a proper location of the road. The intended course must be laid off by the eye, and the upper side of the road marked. If it passes through woods, it will be necessary to use grubbing hoes for the digging. With these, begin at the distance of four or five feet below the marked line, and dig horizontally onward to it. That earth is to be

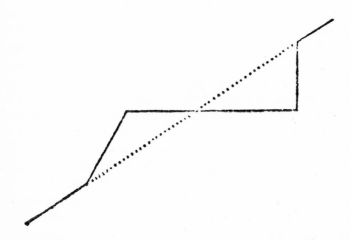

pulled back with broad hoes, and laid over a width of three or four feet below the place from which it was taken. Thus the upper side of the road is formed by cutting down, and the lower side by filling up, with the earth taken from above. After shaping the road roughly, the deficiencies will be seen and may be corrected in the finishing work, by deepening some places and filling up others, so as to graduate the whole properly. A width of eight or nine feet of firm road, will be sufficient for carting marl. If the land through which the road is to be cut is not very steep, and is free from trees and roots, the operation may be made much cheaper by using the plough. The first furrow should be run along the line of the lower side of the intended road, and turned down hill: the plough then returns empty, to carry a second furrow by the first. In this manner it proceeds — cutting deeply, and throwing the slices far, (both of which is easily done on a hill-side,) until rather more than the required width is ploughed. The ploughman then begins again over his first furrow, and ploughs the whole over as at first — and this course is repeated perhaps once or twice more, until enough earth is cut from the upper and put on the lower side of the road. After the first ploughing, eight or ten broad hoes should aid and complete the work, by pulling down the earth from the high to the low side, and particularly in those places where the hill-side is steepest. After the proper shape is given, carts at first empty, and then with light loads, should be driven over every part of the surface of the road, until it is firm. If a heavy rain should fall before it has been thus trodden, the road would be rendered useless for a considerable time.

Tumbril carts drawn by a single horse or mule, are most convenient for conveying marl short distances. Every part of the cart should be light, and the body should be so small as only to hold the load it is intended to carry, without a tail-board. This plan enables the drivers to measure every load, which advantage will be found on trial much more important than would at first be supposed. If carts of common size are used, the careless labourers will generally load too lightly — yet sometimes will injure the horse by putting in a load much too heavy. The small-sized cart-bodies prevent both these faults. The load cannot be made much too heavy —

and if too light, the farmer can detect it at a glance. Where there is a hill to ascend, five heaped bushels of wet marl is a sufficient load for a horse. If the marl is dry, or the road level, six bushels may be put in the same carts, by using tail-boards.

Strong labourers are required in the pit for digging and loading: but boys who are too small for any other regular farm labour, are sufficient to drive the carts. Horses or mules kept at this work soon become so tractable, that very little strength or skill is required to drive them.

All these hints and expedients, or perhaps better plans, would occur to most persons before they are long engaged in marling. Still these directions may help to smooth the obstructions in the way of the inexperienced — and they will not be entirely useless, if they serve to prevent even small losses of time and labour.

My task is at last completed. Whether I shall be able to persuade my countrymen to prize the treasures, and seize the profits which are within their reach, or whether my testimony and arguments shall be fruitless, soon or late, a time must arrive when my expectations will be realized. The use of calcareous manures is destined to change a large portion of the soil of Lower Virginia from barrenness to fertility — which, added to the advantages we already possess — our navigable waters and convenient markets, the facility of tilling our lands, and the choice of crops offered by our climate — will all concur to increase ten-fold the present value of our land, and produce more farming profit than has been found elsewhere on soils far more favoured by nature. Population, wealth, and learning, will keep pace with the improvement of the soil — and we, or our children, will have reason to rejoice, not only as farmers, but as Virginians, and as patriots.

APPENDIX

[A. Page 8.]

Different significations of calcareous earth.

THE definition of calcareous earth, which confines that term to the carbonate of lime, is certainly liable to objections, but less so than any other mode of arrangement. It may at first seem absurd to consider as one of the three earths which compose soils, *one* only of the many combinations of lime, rather than either pure lime alone, or lime in all its combinations. One or the other of these significations is adopted by the highest authorities, when the calcareous ingredients of soil are described — and in either sense, the use of this term is more conformable with scientific arrangement, than mine. Yet much inconvenience is caused by thus applying the term calcareous earth. If applied to *lime*, it is to a substance which is never found existing naturally, and which will always be considered by most persons as the product of the artificial process of calcination, and as having no more part in the composition of natural soils, than the manures obtained from oilcake, or pounded bones. It is equally improper to include under the same general term the combinations of lime, with all the fifty or sixty various acids. Two of these, the sulphate, and the phosphate of lime, are known as valuable manures; but they exist naturally in soils in such minute quantities, and so rarely, as not to deserve to be considered as important ingredients. A subsequent part of this essay will show why the oxalate of lime is also supposed to be highly valuable as a manure, and far more abundant. Many other salts of lime are known to chemists: but their several qualities, as affecting soils, are entirely unknown — and their quantities are too small, and their presence too rare, to require consideration. If all these different combinations of lime, with various and unknown properties, had not been excluded by my definition of calcareous earth, continual exceptions would have been necessary, to avoid stating what was not meant. The carbonate of lime, to which I have confined that term, though only one of many existing combinations, yet in quantity and in importance, as an ingredient of soils, as well as of the known portion of the globe, very far exceeds all the others.

But even if calcareous earth, as defined and limited, is admitted to be the substance which it is proper to consider as one of the three earths of agriculture, still there are objections to its name, which I would gladly avoid. However strictly defined, many readers will attach to terms such meanings as they had previously understood: and the word calcareous has been so loosely, and so differently applied in common language and in agriculture, that much confusion may attend its use. Any thing "partaking of the nature of lime" is "calcareous," according to Walker's Dictionary: Lord Kames limits the term to *pure lime* [1] — Davy [2] and Sinclair,[3] include under it pure lime and all its combinations — and Kirwan,[4] Rozier,[5] and Young,[6] whose example I have followed, confine the name calcareous earth to the carbonate of lime. Nor can any other term be substituted without producing other difficulties. *Carbonate of lime* would be precise, and it means exactly the same chemical substance: but there are insuperable objections to the frequent use of chemical names in a work addressed to ordinary readers. Chalk, or shells, or mild lime, (or what had been quicklime, but which from exposure to the air, had again become carbonated,) all these are the same chemical

[1] *Gentleman Farmer* (2nd ed., Edinburgh [1779]), p. 264.
[2] *Agr. Chem.* (Philadelphia, 1821), p. 223.
[3] [Sir John Sinclair] *Code of Agriculture* (Hartford, 1818), p. 134.
[4] [Richard] Kirwan, *Manures* [London, 1796], chap. 1.
[5] "Terres" [in Rozier, ed.], *Cours Complet d'Agriculture* [13 vols., Paris, 1785–1809].
[6] [Arthur Young] *Essay on Manures* [London, 1804], chap. 3.

substance — but none of these names would serve, because each would be supposed to mean such certain form or appearance of calcareous earth, as they usually express. If I could hope to revive an obsolete term, and with some modification establish its use for this purpose, I would call this earth *calx*, and from it derive *calxing*, to signify the application of calcareous earth, in any form, as manure. A general and definite term for this operation is much wanting. Liming, marling, applying drawn ashes, or the rubbish of old buildings, chalk, or limestone gravel — all these operations are in part, and some of them entirely, that manuring that I would thus call *calxing*. But because their names are different, so are their effects generally considered — not only in those respects where differences really exist, but in those where they are precisely alike.

[B. Page 13.]

The names usually given to soils often incorrect.

Nothing is more wanting in the science of agriculture, than a correct nomenclature of soils, by which the characters might be learned from the names — and nothing has hitherto seemed less attainable. The modes of classing and naming soils, used by scientific authors, are not only different, and opposed to each other — but each one of them is quite unfit to serve the purpose intended. As to the crowd of inferior writers, it is enough to say that their terms are not fixed by any rule — convey no precise meaning, and are worth not much more than those in common use among ourselves, and other practical cultivators, which often vary in their meaning within forty miles of distance. To enable us to judge of the fitness of the names given to soils by others, let us examine those applied by ourselves. We generally describe soils by making a mental comparison with those we are most ac-

customed to — and though such a description is understood
well enough through a particular district, it may have quite
a different meaning elsewhere. What are called *clay* or *stiff*
soils in Sussex and Southampton, would be considered *sandy*
or *light* soils in Goochland — merely because almost every
acre of land in the former counties is sandy, and in the latter,
clays are nearly as abundant.

The conflict of definitions, and consequent confusion in
terms, cannot be more plainly set forth, than by quoting from
some of the highest authorities, the various and contradictory
explanations of a term so common that it is used by every
one who writes or speaks of soils.

"*Loam* denotes any soil moderately cohesive, and more so
than loose chalk. By the author of the Body of Agriculture,
it is said to be a *clay mixed with sand.*" [*Kirwan*, Manures —
Chap. 1.]

"*Loam*, or that species of *artificial soil*, into which the others
are generally brought by the course of long cultivation." —
"Where a soil is moderately cohesive, less tenacious than
clay, and more so than sand, it is known by the name of loam.
From its frequency, there is reason to suppose that *in some
cases* it might be called an *original soil.*" [*Sinclair's* Code of
Agriculture — *Chap.* 1.]

"The word loam should be limited to soils containing at
least one-third of impalpable earthy matter, *copiously
effervescing with acids.*" [*Davy's* Agricultural Chemistry —
Lecture 4.]

"By loam is meant *any of the earths combined with decayed
animal or vegetable matter.*" [*Appendix to Agr. Chem. by
George Sinclair.*] [7]

"Loam — *fat unctuous earth — marl.*" [*Johnson's Dictionary, 8vo. Ed., and also Walker's.*] [8]

---

[7] [Probably George Sinclair, *Hortus Gramineus Woburnensis* (London,
1816).]

[8] [Probably Samuel Johnson, *A Dictionary of the English Language*,

"Loam may be considered a clay of loose or friable consistency, mixed with *mica* or isinglass, and *iron ochre*."
[American Farmer, III (1821), 320.]

[C. Page 20.]

Some effects of slavery on agricultural profits.

The cultivators of Lower Virginia derive a portion of their income from a source distinct from tillage — and which, though it often forces them to persist in their profitless farming, yet also conceals, and mitigates its consequences. This source of income is the breeding and selling of slaves. I do not mean to charge any person with intentionally carrying on a regular business of this kind: but whether we wish it or not, from the nature of existing circumstances, we all are acting some part in aid of a general system, which taken altogether, is precisely what I have named. No man is so inhuman as to breed and raise slaves, to sell off a certain proportion regularly, as a western drover does with his herds of cattle. But sooner or later the general result is the same. Sales may be made voluntarily, or by the sheriff — they may be met by the first owner, or delayed until the succession of his heirs — or the misfortune of being sold may fall on one parcel of slaves, instead of another: but all these are but different ways of arriving at the same general and inevitable result. With plenty of wholesome, though coarse food, and under such mild treatment as our slaves usually experience, they have every inducement and facility to increase their numbers with all possible rapidity, without any opposing check, either prudential, moral, or physical. These several checks to the increase of population operate more or less on all free persons, whether rich or poor — and slaves, situated as ours are, perhaps are placed in the

2 vols. (London, 1755), and John Walker, *A Critical Pronouncing Dictionary* . . . (London, 1791).]

only possible circumstances, in which no restraint whatever prevents the propagation and increase of the race. From the general existence of this state of circumstances, the particular effects may be naturally deduced: and facts completely accord with what those circumstances promise. A gang of slaves on a farm will often increase to four times their original number, in less than forty years. If a farmer is only able to feed and maintain his slaves, their increase in value may double the whole of his capital originally vested in farming, before he closes the term of an ordinary life. But few farms are able to support this increasing demand, and furnish the necessary supplies to the family of the owner — whence very many owners of large estates in lands and negroes, are throughout their lives too poor to enjoy the comforts of wealth, or to encounter the expenses necessary to improve their unprofitable farming. A man so situated, may be said to be a slave to those, whom the law has made slaves to himself. If the slave owner is industrious and frugal, he may be able to support the increasing number of his slaves, and to bequeath them undiminished to his children. But the income of few persons increases as fast as their slaves — and if not, the consequence must be, that some of them will be sold, that the others may be supported; and the sale of more is perhaps afterwards compelled, to pay debts incurred in striving to put off that dreaded alternative. The slave first almost starves his master, and at last, is eaten by him — at least he is exchanged for his value in food. The sale of slaves is always a severe trial to their owner; obstacles are opposed to it, not only sentiments of humanity, and of regard for those who have passed their lives in his service — but every feeling he has of false shame comes to aid — and such sales are generally postponed, until compelled by creditors, and are carried into effect by the sheriff, or by the administrator of the debtor. And when the sale finally takes place, its magnitude makes up for all previous delays. Do what we will, the surplus slaves *must* be sent out of a country which is not able to feed them: and these causes continue to supply the immense numbers that are annually sold and carried away from Lower Virginia, without producing the political benefit, of lessening the actual number remaining. Nothing can check this forced emigration of blacks, and the voluntary emigration of whites, except increased production of food, obtained by enriching our lands, and the consequent increase of farming profits. No effect will more certainly follow its cause than this — that whenever our land is so improved as to produce double its present supply of food, it will also have, and will retain, double its present amount of population. The improving farmer who adds one hundred bushels of corn to the previous product of his country, effectually adds also to its population, as many persons as his increase of product will feed.

[D. Page 36.]

Opinions that soils are generally calcareous.

I have asserted that the inference to be drawn from all the descriptions of soils, in the most esteemed treatises on agriculture, is that *calcareous earth* is a very general, if not a universal ingredient. This assertion can be proved beyond all doubt, from European authors: but it would require many and long extracts, too bulky to include here, and which cannot be fairly abridged, or exhibited by a few examples. No author says directly that calcareous earth is present in all soils — but its being always named as one of the ingredients of soils in general, and no cases of its absolute deficiency being directly stated, amount to the declaration that calcareous earth is very rarely, if ever, entirely wanting in any soil. We may find enough directions to apply calcareous manures to soils that are deficient in that ingredient: but that deficiency, seems to be not spoken of as *absolute*, but *relative* to other soils more abundantly supplied. In the same manner, they direct clay, or sand, to be used as manure for soils very deficient in one or the other of those earths — but without meaning that any soil under cultivation can be found entirely destitute of sand, or of clay. My proofs from general treatises, would therefore be generally indirect — and the quotations necessary to exhibit them, would show what had *not* been said, rather than what *had* — that they did *not* assert the absence of calcareous earth, instead of directly asserting its universal presence. Extracts for this purpose, however satisfactory, would necessarily be too voluminous, and it is well that they can be dispensed with. Better proof, because it is direct, and more concise, will be furnished by quoting the opinions of a few agriculturists of our own country, who are extensively acquainted with European authors, and have evidently drawn their opinions from those sources. These quotations will not only show conclusively, that their authors consider the received European doctrine to be that all soils are more or less calcareous — but also, that they apply the same general character to the soils of the United States, without expressing a doubt, or naming an exception.

1st. From a "Treatise on Agriculture," (ascribed to General Armstrong,) published in the *American Farmer*, [I. (1819), 153.]

"Of six or eight substances, which chymists have denominated earths, four are *widely and abundantly diffused*, and form the crust of our globe. These are silica, alumina, lime, and magnesia." — "In a pure or isolated state, these earths are wholly unproductive; but when decomposed and mixed, and to this mixture is added the residuum of dead animal or vegetable matter, they become fertile, and take the general name of soils, and are again denominated, after the earth that most abounds in their composition respectively.—"

2. Address of R. H. Rose to the Agricultural Society of Susquehanna. [*Am. Far.* III, 101.]

"Geologists suppose our earth to have been masses of rock of various kinds, but principally silicious, aluminous, calcareous, and magnesian — from the gradual attrition, decay, and *mixture* of which, together with an addition of vegetable and animal matter, is formed the soil; and this is called sandy, clayey, calcareous, or magnesian, according as the particular primitive material, preponderates in its formation."

3. Address of Robert Smith to the Maryland Agricultural Society. [*Am. Far.* III, 228.]

"— The soils *of our country* are in general clay, sand, gravel, clayey loam, sandy loam, and gravelly loam. Clay, sand, and gravel, need no description, &c." — "*Clayey loam* is a compound soil, consisting of clay, and sand or gravel, with a mixture of *calcareous matter*, and in which clay is predominant. — *Sandy, or gravelly loam*, is a compound soil, consisting of sand or gravel, and clay, with a mixture of *calcareous matter*, and in which sand or gravel is predominant."

The two first extracts merely state the geological theory of the formation of soils, which is received as correct by the most eminent agriculturists of Europe. How far it may be supported or opposed by the actual constitution, and number of ingredients, of European soils, is not for me to decide on, nor is the consideration necessary to my subject. But the adoption of this general theory, by American writers, without excepting American soils, is an indirect, but complete application to them, of the same character and composition. The writer last quoted, states positively that the various loams, (which comprise at least nineteen twentieths of *our* soils, and I presume also of the soils of Maryland,) contain calcareous matter. The expression of this opinion by Mr. Smith, is sufficient to prove that such was the fair and plain deduction from his general reading on agriculture, from which source only could his opinions have been derived. If the soils of Maryland are not very unlike those of Virginia, I will venture to assert, that not one hundredth part of all the clayey, sandy, and gravelly loams, contains the smallest proportion of *carbonate of lime* — and that not a single specimen of calcareous soil can be found, between the falls of the rivers, and the most eastern body of limestone.

But though the direct testimony of European authors, (as cited in the essay,) concurs with the indirect proofs referred to in this note, to induce the belief that soils are very rarely destitute of calcareous earth, yet statements may be found of some particular soils, being considered of that character. These statements, even if presented by the authors of general trea-tises, would only seem to present exceptions to their general rule of the almost universal diffusion of calcareous earth in soil. But so far as I know, no such exceptions are named in the description of soils in any general treatise, and therefore have not the slightest effect in contradicting or modifying their testimony on this subject. It is in the description of soils of particular farms, or districts, that some such statements are made: and if no such examples had been mentioned, they would not have been needed to prove the existence in Europe of soils like most of ours, destitute of calcareous earth. These facts do not oppose my argument. I have not asserted, (nor believed, since I have endeavoured to investigate this subject,) that there were not soils, and perhaps many extensive districts, containing no calcareous earth. My argument merely main-tains that these facts would not be inferred, but the contrary, by any general and cursory reader of the agricultural treatises of Europe, that we are best acquainted with. It has not been my purpose to inquire as to the existence, or the extent of soils of this kind in Europe. But judging from the indirect testimony furnished by accounts of the minerals, and vege-table productions in general descriptions of different countries, I would suppose that soils having no calcareous earth were often found in Scotland and the northern part of Germany, and that they were comparatively rare, in England and France.

[E. Page 72.]

Calcareous earth a preserver of putrescent animal matter.

My experiment of the putrefaction of animal matter in contact with calcareous earth, was commenced with a view to results different from what were obtained, as stated in the essay. Darwin [9] says that nitrous acid is produced in the process of putrefaction, and he supposes the nitrate of lime to be very serviceable to vegetation. [*Phytologia, p. 210 and 224, Am. Ed.*] As the nitrous acid is a gas, it must pass off into the air, under ordinary circumstances, as fast as it is formed, and be entirely lost. But as it is strongly attracted by lime, a cover of calcareous earth ought to arrest it, and form a new combination, which, if not precisely nitrate of lime, would at least be composed of the same elements, though in different proportions. To ascertain whether any such combination had taken place, when the manure was used, I took a handful of the calcareous earth, which had been in immediate contact with the carcass, and threw it into a glass of hot water. After remaining half an hour, the fluid was poured off, filtered and evaporated, and left a considerable proportion of a white soluble salt (supposed eight or ten grains.) I could not ascertain its kind — but it was not deliquescent, and therefore could not have been the nitrate of lime. The spot on which the carcass laid, was so strongly impregnated by this salt, that it remained bare of vegetation for several years.

But whatever were the products of fermentation saved by this experiment, it was sufficiently evident that little or nothing was lost; as every thing is, when flesh putrefies in the open air. And I presume that a cover of a few inches, or even a few feet, of aluminous or silicious earth, or both, would have

---

[9] [Probably Erasmus Darwin (1731–1802), English physician, scientist, poet, who wrote *Phytologia; or the Philosophy of Agriculture, and Gardening* (London, 1800).]

very little effect, in preserving any aëriform products of putrefaction. Whenever the carcasses of animals, or any other
animal substances subject to rapid and wasteful fermentation,
can be obtained in great quantity, their enriching power might
be all secured by depositing them between layers of fossil
shells, or any other form of calcareous earth. I have understood that on the borders of the Chowan, immense quantities
of herrings are often used as manure, as purchasers cannot
always take off the myriads with which they are supplied. A
herring is buried under every corn-hill, and fine crops are
thus made, as far as this singular mode of manuring can be
extended. But whatever benefits may be thus derived, the
sense of smelling, as well as the known chemical process of
putrefaction, make it certain that nine tenths of all this rich
manure must be wasted in the air. If those who fortunately
possess this supply of animal manure, would let the fermentation take place and be completed, mixed with and enclosed
by calcareous earth, in pits of suitable size, they would increase
prodigiously, both the amount and the permanency of their
acting animal manure, besides obtaining the benefit of the
calcareous manure mixed with it. This opinion is principally
founded on theory, or upon the received opinions of the
chemical properties and affinities of the different substances
and their products. My farm, like most others, furnishes no
considerable or regular supply of animal substances requiring
such care to preserve, and therefore, my practice in this respect has been very limited.

[F. Page 80.]

Marling in England. — Liming.

Custom compels me to use the name *marl* for our deposits
of fossil shells. But as I have defined the manuring by this
substance, which is called *marling*, to be simply *making a soil*

calcareous, or more so than it was before, any term used for that operation would serve, if its meaning was always kept in view. But this unfortunately is of old and frequent use in English books, with very different meanings. These differences have been generally stated in the body of the essay, and I shall here present the proofs. The following quotations will show that the term *marl* is frequently applied in Britain, to clays containing no known or certain proportion of calcareous earth — that when calcareous earth is known to be contained, it is seldom relied on as the most valuable part of the manure — and that in most cases the reader is left in doubt whether the manure has served to increase or diminish, or has not altered materially, the former calcareous ingredient of the soil.

1. Kirwan, [*Essay on Manures, page 4*] on the authority of Arthur Young, and the Bath Memoirs,[10] states "that in some parts of England, where husbandry is successfully practised, any loose clay is called marl: in others, marl is called chalk, and in others, clay is called loam."

2. The learned and practical Miller[11] thus defines and describes marl, in *The Abridgment of the Gardener's Dictionary*, fifth London edition, at the article "Marl."

"Marl is a kind of clay which is become fatter and of a more enriching quality, by a better fermentation, and by its having lain so deep in the earth as not to have spent or weakened its fertilizing quality by any product."

"Marls are of different qualities in different counties of England." He then names and describes ten varieties, most of them being very minutely and particularly characterized — and in only two of the ten, is there any allusion to a calcareous ingredient, and in these, it is evidently not deemed to constitute

---

[10] [Probably Arthur Young, *Essay on Manures*, one edition of which was published as *Bath and West of England Society of Agricultural Letters*, X (1805), 197–198.]

[11] [Probably Philip Miller (1691–1771), English gardener and botanical writer, whose *The Gardener's Dictionary*, 2 vols. (London, 1731–1739), went through many editions.]

their value as manures. These are "the cowshut marl" of Cheshire, which "is of a brownish colour, with blue veins in it, and little lumps of chalk or limestone" — and "clay-marl; this resembles clay, and is pretty near akin to it, but is fatter, and sometimes mixed with chalk stones."

"The properties of any sorts of marls, by which the goodness of them may be best known, are better judged of by their purity and uncompoundedness, than their colour: as if it will break in pieces like dice, or into thin flakes, or is smooth like lead ore, and is without a mixture of gravel or sand; if it will shake like slatestones, and shatter after wet, or will tumble into dust, when it has been exposed to the sun; or will not hang and stick together when it is thoroughly dry, like tough clay; but is fat and tender, and will open the land it is laid on, and not bind; it may be taken for granted that it will be beneficial to it."

3. Johnson's *Dictionary* (Octavo edition) defines marl in precisely the words of the first sentence of Miller, as quoted above.

4. Walker's *Dictionary* (Octavo edition) gives only the following definition — "Marl — a kind of clay much used for manure."

5. *A Practical Treatise on Husbandry*,[12] (2d London Ed. 4to, 1762,) which professes to be principally compiled from the writings of Duhamel, Evelyn, Home, and Miller,[13] supplies the following quotations.

Page 27. "But of all the manures for sandy soils, none is so good as marl. There are many different kinds and colours of it, severally distinguished by many writers; but their virtue is the same; they may be all used upon the same ground, without the smallest difference in their effect. The colour is either red, brown, yellow, grey, or mixed. It is to be known by its

[12] [Probably Henri L. Duhamel Du Monceau, *A Practical Treatise of Husbandry* . . . (London, 1762).]

[13] [John Evelyn (1620–1706), Henry Home (1696–1782), and Philip Miller (1691–1771), all English agricultural writers.]

pure and uncompounded nature. There are many marks to distinguish it by; such as its breaking into little square bits; its falling easily into pieces, by the force of a blow, or upon being exposed to the sun and the frost; its feeling fat and oily, and shining when 'tis dry. But the most unerring way to judge of marl, and know it from any other substance, is to break a piece as big as a nutmeg, and when it is quite dry, drop it into a glass of clear water, where, if it be right, it will dissolve and crumble, as it were, to dust, in a little time, shooting up sparkles to the surface of the water." — Not the slightest hint is here of any calcareous ingredient being necessary to, or even serving in any manner to distinguish marl. But afterwards, in another part of this work, (page 29) when *shell marl* is slightly noticed, it is said, "this effervesces strongly with all acids, which is perhaps chiefly owing to the shells. There are very good marls which show nothing of this effervescence: and therefore the author of the *New System of Agriculture* judged right in making its solution in water the distinguishing mark."

The last sentence declares, as clearly as any words could do, that no calcareous ingredient is necessary, either to constitute the character, or the value of marl. And though it may be gathered from other parts of this work, that what is called marl generally contains calcareous earth, yet no importance seems attached to that quality, any more than to the particular colour of the earth, or any other accidental or immaterial appearance of some of the varieties described.

The "shell marl" alluded to above, without explanation might be supposed to be similar to our beds of fossil shells, which are called marl. The two manures are very different in form, appearance, and value, though agreeing in both being calcareous. The manure called shell marl by the work last quoted from, is described there with sufficient precision, and more fully in several parts of the Edinburgh Farmer's Magazine, and in the Memoirs of the Philadelphia Agricultural Society, [*Vol.* 3. *page* 206.] It is still more unlike *marl* properly

so called, or any of the substances described under that name
alone, in the foregoing, or following quotations. This manure
is almost a pure calcareous earth, being formed of the remains
of small fresh-water shells deposited on what were once the
bottoms of lakes, but which have since become covered with
bog or *peat* soil. If I may judge from our beds of muscle shells,
(to which this manure seems to bear most resemblance,) much
putrescent animal matter is combined with, and serves to give
additional value to, these bodies of shells. This kind of manure
is sold in Scotland by the bushel, at such prices, as show that
its value is well understood. It seems to be found but in few
situations, and though called a kind of marl, is never meant
when that term only is used generally.

The opinions expressed in the foregoing extracts, prove
sufficiently that it was not the ignorant cultivators only, who
either did not know of, or attached no importance to the
calcareous ingredient in marl: and it was impossible, that from
any number of such authors, an American reader could learn
that either the object, or the effect of *marling*, was to render a
soil more calcareous — or that our bodies of fossil shells
resembled marl in character, or in operation, as a manure. Of
this, the following quotation will furnish striking proof —
and the more so, as the author refers frequently to the works
of Anderson, and of Young, who treated of marl and calcareous manures, in a more scientific manner than had been usual.
This author, Bordley, cannot be justly charged with inatten-
tion to the instruction to be gained from books: for his greatest
fault, as an agriculturist, is his fondness for applying the practices of the most improved husbandry of England, to our
lands and situations, however different and unsuitable — which
he carries to an extent that is ridiculous as theory, and would
be ruinous to the farmer who should so shape his general
practice.

6. Bordley's *Husbandry*,[14] 2d edition. [*Note to page 55.*]

[14] [Probably John B. Bordley (1727–1804), American agricultural writer,

"I farmed in a country [the Eastern Shore of Maryland] where habits are against a due attention to manures: but having read of the application of marl as a manure, I inquired where there was any in the peninsula of the Chesapeake, *in vain.* My own farm had a greyish clay, which to the eye was marl: but because it did not effervesce with acids, it was given up, when it ought to have been tried on the land, especially as it rapidly crumbled and fell to mud, in water, with some appearance of effervescence." — That peninsula, through which Mr. Bordley in vain inquired for marl, has immense quantities of the fossil shells which we so improperly call by that name. But as his search was directed to *marl* as described by English authors — and not to calcareous earth simply — it is not to be wondered at that he should neither find the one substance, or attach enough importance to the other, to induce the slightest remark on its probable use as manure.

7. The *Practical Treatise on Husbandry*, page 21, among the directions for improving clay land, has what follows. "Sea sand and sea shells are used to great advantage as a manure, chiefly for cold strong [i. e. clay,] land, and loam inclining to clay. They separate the parts; and the *salts* which are contained in them are a very great improvement to the land. Coral, and such kind of stony plants which grow on the rocks, are filled with salts, which are very beneficial to land. But as these bodies are hard, the improvement is not the first or second year after they are laid on the ground, because they require time to pulverize them, before their salts can mix with the earth to impregnate it. The consequence of this is, that their manure is lasting. Sand, and the smaller kind of sea weeds, will enrich land for six or seven years: and shells, coral, and other hard bodies, will continue many years longer.

In some countries *fossil shells* have been used with success as manure; but they are not near so full of salts, as those shells

<hr>

author of *Essays and Notes on Husbandry and Rural Affairs* (Philadelphia, 1799).]

which are taken from the sea shore; and therefore the latter are always to be preferred. Sea sand is much used as manure in Cornwall. The best is that which is intimately mixed with coral." After stating the manner in which this "excellent manure" is taken up from the bottom, in barges, its character is thus continued: "It [i. e. the sea sand mixed with coral, as it may happen,] gives the heat of lime, and the fatness of oil, to the land it is laid upon. Being more solid than shells, it conveys a greater quantity of fermenting earth in equal space. Besides, it does not dissolve in the ground so soon as shells, but decaying more gradually, continues longer to impart its warmth to the juices of the earth."

Here are described manures which are known to be calcareous, which are strongly recommended — but solely for their supposed mechanical effect in separating the parts of close clays, and on account of the salts derived from sea water, which they contain. Indeed, no allusion is made to any supposed value, or even to the presence of calcareous earth, which forms so large a proportion of these manures: and the fossil shells, (in which that ingredient is more abundant, more finely reduced, and consequently more fit for both immediate and durable effects,) are considered as less efficacious than solid sea shells — and inferior to sea sand. All these substances, besides whatever service their salts may render, are precisely the same kind of calcareous manure, as our beds of fossil shells furnish in a different form. Yet neither here nor elsewhere, does the author intimate that these manures and marl have similar powers for improving soils.

The foregoing quotations show what opinions have been expressed by English writers of reputation — and what opinion would be formed by a general reader of these and other agricultural works, of the nature of what is called marl, in England, as well as what is so named in this part of our country. I do not mean that other authors have not thought more correctly, and sometimes expressed themselves with precision on this subject. Mineralogists define *marl*, to be a *calcareous clay* [15] — and in this correct sense, the term is used by Davy, and other chemical agriculturists. Such authors as Young, and Sinclair, also could not have been ignorant of the true composition of marl — yet even they have used so little precision or clearness when speaking of the effects of marling, that their statements, (however correct they may be in the sense they intended them,) convey no exact information, and have not served to remove the erroneous impressions made by the great body of their predecessors. Knowing as Young did [see first quotation] the confusion in which this subject was involved, it was the more incumbent on him to be guarded in his use of terms so generally misapplied. Yet considering his practical and scientific knowledge as an agriculturist, his extensive personal observations, and the quantity of matter he has published on soils and calcareous manures, his omissions are more remarkable than those of any other writer. In such of his works as I have met with, though full of strong recommendations of marling, in no case does he state the composition of the soil, (as respects its calcareous ingredient,) or the proportion added by the operation — and generally notices neither, as if he viewed marling just as most others have done. These charges are supported by the following extracts and references.

8. Young's *Farmer's Calendar*, 10th London edition, page 40. — On marling. Through nearly four pages this practice is strongly recommended — but the manures spoken of, are regularly called "marl or clay," and their application, "marling or claying." Mr. Rodwell's account of his practice is inserted at length. On leased land he "clayed or marled" eight hundred and twenty acres with one hundred and forty thousand loads, and at a cost of four thousand nine hundred and fifty-eight pounds — and the business is stated to have been attended with great profit. At last, the author lets us know that it is not the same substance that he has been calling "marl or clay"

[15] [Parker] Cleaveland, [*An Elementary Treatise on*] *Mineralogy* [*and Geology* (Boston, 1816)].

— and that the marl effervesces strongly with acids, and the clay slightly. But we are told nothing more precise as to the amount of calcareous ingredients, either in the manures, or the soil — and even if we were informed on those heads, (without which we can know little or nothing of what the operation really is,) we are left ignorant of how much was clayed, and how much marled. It is to be inferred, however, that the clay was thought most serviceable, as Mr. Rodwell says "clay is much to be preferred to marl on those sandy soils, some of which are loose, poor, and even a black sand."

9. Young's *Survey of Norfolk*,[16] (a large and closely printed octavo volume,) has fourteen pages filled with a minute description of the soils of that county — but without any indication whatever of the proportion, presence, or absence, of calcareous earth in that extensive district of sandy soils, so celebrated for their improvement by marling — nor in any other part of the county. The wastes are very extensive: one of them (page 385) eighteen miles across, quite a desert of sand, "yet highly improvable." Of this also, no information is given as to its calcareous constitution.

10. The section on marl (page 402, of the same work) gives concise statements of its application, with general notices of its effects, on near fifty different parishes, neighbourhoods, or separate farms. Among all these, the only statements from which the calcareous nature of the manure may be gathered, are, (page 406) of a marl that "ferments strongly with acids" — another, (page 409,) that marling at a particular place destroys sorrel — and (page 410) that the marl is generally calcareous, and that that containing the most clay, and the *least* calcareous earth, is preferred by most persons, but not by all.

Young's *General View of the Agriculture of Suffolk*,[17] (an

[16] [Arthur Young, *General View of the Agriculture of Norfolk* (London, 1804).]
[17] [*General View of the Agriculture of the County of Suffolk* (London, 1794).]

octavo of 432 pages of close print,) in the description of soils, affords no information as to any of them being calcareous, or otherwise: yet the author mentions (page 3) having analyzed some of the soils, and reports their aluminous and silicious ingredients. Nor can more be learned, in this respect, in the long account afterwards given of the "marl" which has been very extensively applied also in the county of Suffolk. We may gather however from the following extracts, that the "marl or clay" of Suffolk, is generally calcareous, but that this quality is not considered the principal cause of its value — and further, that *crag*, a much richer calcareous manure, (which seems to be the same with our richest beds of fossil shells, or marl,) is held to be injurious to the sandy soils, which are so generally improved by what is there called marl.

11. Page 186. — "Claying — a term in Suffolk, which includes marling; and indeed the earth carried under this term is very generally a clay marl; though a pure, or nearly a pure clay, is preferred for very loose sands."

12. Page 187. — After speaking of the great value of this manure on light lands, he adds — "But when the clay is not of a good sort, that is, when there is really none, or scarcely any clay in it, but is an imperfect and even a hard chalk, there are great doubts how far it answers, and in some cases has been spread to little profit."

13. Page 5. — "Part of the under stratum of the county is a singular body of cockle and other shells, found in great masses in various parts of the country, from Dunwich quite to the river Orwell, &c." "I have seen pits of it to the depth of fifteen or twenty feet, from which great quantities had been taken for the purpose of improving the heaths. It is both red and white, and the shells so broken as to resemble sand. On lands long in tillage, *the use is discontinued*, as it is found to make the sands blow more." [*That is, to be moved by the winds.*]

14. Sinclair's *Code of Agriculture*,[18] p. 138. — "*Marl.* Of this substance, there are four sorts, rock — slate — clay — and shell marl. The three former are of so heavy a nature that they are seldom conveyed to any distance; though useful when found below a *lighter* soil. But shell marl is specifically lighter, and consists entirely of calcareous matter, (the broken and partially decayed shells of fish,) which may be applied as a top dressing to wheat and grass, when it would be less advantageous to use quicklime." [This is the kind of manure referred to in extract 5, and there more particularly described.] "In Lancashire and Cheshire, clay, or red marl, is the great source of fertilization, &c." "The quantity used is enormous; in many cases about three hundred middling cart loads per acre, and the fields are sometimes so thickly covered as to have the appearance of a red soiled fallow, fresh ploughed." This account of the Lancashire improvements made by red clay marl, closes with the statement that "the effects are represented to be beneficial in the highest degree" — which is fully as exact an account of profit, or increased production, as we can obtain of any other marling. Throughout, there is no hint as to the calcareous constituents of the soil or the manure, or whether either rock, clay, or slate marls generally, are valuable for that, or for other reasons; nor indeed could we guess that they contained any calcareous earth, but for their being classed, with many other substances, under the general head of calcareous manures.

15. *Code of Agriculture*, page 19. — "The means of ameliorating the texture of *chalky* soils, are either by the application of clayey and sandy loams, pure clay, or *marl*." "The chalk stratum sometimes lies upon a thick vein of black tenacious marl, of a rich quality, which ought to be dug up and mixed with the chalk."

[18] [Sir John] Sinclair, *The Code of Agriculture* . . . [London, 1817], p. 138.

16. Dickson's Farmer's Companion.[19] — The author recommends "argillaceous marl" for the improvement of chalky soils: and for sandy soils, "where the calcareous principle is in sufficient abundance, argillaceous marl, and clayey loams," are recommended as manures.

17. Kirwan on Manures, page 80. — "*Chalky loam.* The best manure for this soil is clay, or argillaceous marl, if clay cannot be had; because this soil is defective principally in the argillaceous ingredient."

The evident intention and effect of the marling recommended in the three last extracts, is to diminish the proportion of calcareous earth in the soil.

18. Kirwan on Manures, page 87. — "In Norfolk, they seem to value clay more than marl, probably because their sandy soils already contain calcareous parts." From this it would follow, that the great and celebrated improvements in Norfolk, made by marling, had actually operated to lessen the calcareous proportion of the soil, instead of increasing it. Or if so scientific and diligent an inquirer as Kirwan was deceived on this very important point, it furnishes additional proof of the impossibility of drawing correct conclusions on this subject from European books — when it is left doubtful, whether the most extensive, the most profitable, and the most celebrated improvements by "marling," in Europe, have in fact served to make the soil more or less calcareous.

Most of the extracts which I have presented, are from British agriculturists of high character and authority. If such writers as these, while giving long and (in some respects) minute statements of marl, and marling, omit to tell, or leave their readers to doubt, whether the manure or the soil is the most calcareous — or what proportions of calcareous earth, or whether any, is present in either — then have I fully estab-

---

[19] [Probably R. W. Dickson, ed., *The Commercial and Agricultural Magazine* (London, 1799–1816), issued 1807–1812 as *Agricultural Magazine* or *Farmer's Monthly Journal.*]

lished that the American reader who may attempt to draw instruction from such sources, as to the operation, effects and profits of either marl or calcareous manures, will be more apt to be deceived and misled, than enlightened.

Nevertheless, much valuable information may be obtained from these same works, on calcareous manure, or on marl, (in the sense it is used among us) — but under a different head, viz. *lime*. This manure is generally treated of with as little clearness or correctness, as is done with marl: but the reader at least cannot be mistaken in this, that the ultimate effect of every application of lime, must be to make the soil more calcareous — and to that cause solely are to be imputed all the long-continued beneficial consequences, and great profits, which have been derived from liming. But excepting this one point, in which we cannot be misled by ignorance, or want of precision, the mass of writings on lime, as well as calcareous manures in general, will need much sifting to yield instruction. The opinions published on the operation of lime, are so many, so various, and contradictory, that it seems as if each author had hazarded a guess, and added it to a compilation of those of all who had preceded him. For a reader of these publications to be able to reject all that is erroneous in reasoning, and in statements of facts — or inapplicable, on account of difference of soil, or other circumstances — and thus obtain only what is true, and valuable — it would be necessary for him first to understand the subject better than most of those whose opinions he was studying. It was not possible for them to be correct, when treating (as most do) of *lime* as one kind of manure, and every different form of the *carbonate of lime*, as so many others. Only one distinction of this kind (as to operation and effects) should be made, and never lost sight of — and that is one of substance, still more than of name. Pure or quicklime, and carbonate of lime, are manures entirely different in their powers and effects. But it should be remembered that the substance that was *pure lime* when just burned, often becomes *carbonate of lime* before it is used, (by absorbing carbonic acid from the atmosphere,) — still more frequently before a crop is planted — and probably always, before the first crop ripens. Thus, the manure spoken of as lime, is often at first, and always at a later period, neither more nor less than calcareous earth. Lime, which at different periods, is two distinct kinds of manure, is considered as only one: (and to calcareous earth are given as many different names, all considered to have different values and effects, as there are different forms and mixtures of the substance presented by nature.)

Until now, I have been careful to say but little of *pure lime*, for fear of my meaning being mistaken, from the usual practice of confounding it with calcareous earth — or considering its first and later operations, as belonging to one and the same manure. The connexion between the manures is so intimate, yet their actions so distinct, that my subject requires the concise notice of lime which will now be offered.

My own use of lime as a manure has not extended beyond a few acres; and I do not pretend to know any thing from experience, of its first or caustic effects: but Davy's simple and beautiful theory of its operation carries conviction with it, and in accordance with his opinions I shall state the theory; and thence deduce its proper practical use. [*Agr. Chem. Lecture* 7.]

By a sufficient degree of heat, the carbonic acid is driven off from shells, limestone, or chalk, and the remainder is pure or caustic lime. In this state it has a powerful decomposing power on all putrescent animal and vegetable matters, which it exerts on every such substance in the soils to which it is applied as manure. If the lime thus meets with solid and inert vegetable matters, it hastens their decomposition, renders them soluble, and brings them into use and action as manure. But such vegetable and animal matters as well already decomposed, and fit to support growing plants, are injured by the addition of lime — as the chemical action which takes place between these bodies, forms different compounds which are always less valuable than the putrid or soluble matters were, before being acted on by the lime.

This theory will direct us to expect profit from liming all soils containing much unrotted and inert vegetable matter, as our acid woodland when first cleared, and perhaps worn fields, covered with broom grass — and to avoid the application of lime, or (what is the same thing,) to destroy previously its caustic quality by exposure to the air, on all good soils containing soluble vegetable or animal matters, and on all poor soils deficient in inert, as well as active nourishment for plants. The warmth of our climate so much aids the fermentation of all putrescent matters in soils, that it can seldom be required to hasten it by artificial means: to check its rapidity is much more necessary, to avoid the waste of manures in our lands. But in England, and still more in Scotland, the case is very different. There, the coldness and moisture of the climate greatly retard the fermentation of the vegetable matter that falls on the land — so much so, that in certain situations the most favourable to such results, the vegetable cover is increased by the deposit of every successive year, and forms those vegetable soils, which are called *moor*, *peat*, and *bog* lands. Vegetable matter abounds in these soils, sometimes it even forms the greater bulk for many feet in depth — but it is inert, insoluble, and useless, and the soil is unable to bring any useful crop, though containing vegetable matter in such excess. Many millions of acres in Britain, are of the different grades of peat soils, of which not an acre exists in the eastern half of Virginia. Upon this ground of the difference of climate, and its effect on fermentation, I deduce the opinion that *lime* would be serviceable much more generally in Britain than with us: and indeed that there are very few cases in which the caustic quality would not do our arable lands more harm than good. This is no contradiction of the great improvements which have been made on some farms by applying lime — because its caustic quality was seldom allowed to act at all. Lime is continually changing to the carbonate of lime, and in practice, no exact line of separation can be drawn between the transient effects of the one, and the later, but durable improvement from the other. Lime powerfully attracts the carbonic acid of which it was deprived by heat, and that acid is universally diffused through the atmosphere (though in a very small proportion,) and is produced by every decomposing putrescent substance. Consequently caustic lime on land, is continually absorbing and combining with this acid; and with more or less rapidity, according to the manner of its application, is returning to its former state of mild calcareous earth. If spread as a top dressing on grass lands — or on ploughed land, and superficially mixed with the soil by harrowing — or used in composts with fermenting vegetable matter — the lime is probably completely carbonated, before its causticity can act on the soil. In no case can lime, applied properly as manure, long remain caustic in the soil. Thus most applications of lime are simply applications of calcareous earth, but acting with greater power at first, in proportion to its quantity, because more finely divided, and more equally distributed.

[G. Page 118.]

The cause of the inefficacy of gypsum as a manure on acid soils.

I do not pretend to explain the mode of operation by which gypsum produces its almost magic benefits: it would be equally hopeless and ridiculous for one having so little knowledge of the successful practice, to attempt an explanation, in which so many good chemists, and agriculturists both scientific and practical, have completely failed. There is no operation of nature less understood, or of which the cause, or agent, seems so totally disproportioned to the effect, as the enormous increase of vegetable growth from a very small quantity of gypsum, in circumstances favourable to its action. All other known manures, whatever may be the nature of their action, require to be applied in quantities, very far exceeding any bulk of crop expected from their use. But one bushel of gypsum, spread over an acre of land fit for its action, may add more than twenty times its own weight to a single crop of clover.

However wonderful and inscrutable the fertilizing power of this manure may be, and admitting its cause as yet to be hidden, and entirely beyond our reach — still it is possible to show reasons why gypsum cannot act on many situations, where all experience has proved it to be worthless. If this only can be satisfactorily explained, it will remove much of the uncertainty as to the effects to be expected: and the farmer may thence learn on what soils he may hope for benefit from this manure — on what it will certainly be thrown away — and by what means the circumstances adverse to its action may be removed, and its efficacy thereby secured. This is the explanation that I shall attempt.

If the vegetable acid, which I suppose to exist in what I have called acid soils, is not the oxalic, (which is the particular acid in sorrel,) at least every vegetable acid, being composed of different proportions of the same elements, may easily change to any other, and all to the oxalic acid. This, of all bodies known by chemists, has the strongest attraction for lime, and will take it from any other acid which was before combined with it — and for that purpose, the oxalic acid will let go any other earth or metal, which it had before held in combination. Let us then observe what would be the effect of the known chemical action of these substances, on their meeting in soils. If oxalic acid was produced in any soil, its immediate effect would be to unite with its proper proportion of lime, if enough was in the soil in any combination whatever. If the lime was in such small quantity as to leave an excess of oxalic acid, that excess would seize on the other substances in the soil, in the order of their mutual attractive force; and one or more of such substances are always present, as magnesia, or more certainly, iron and alumina. The soil then would not only contain some proportion of the *oxalate of lime*, but also the *oxalate* of either one or more of the other substances named. Let us suppose gypsum to be applied to this soil. This substance, (sulphate of lime) is composed of sulphuric acid and lime. It is applied in a finely pulverized state, and in quantities from half a bushel to two bushels the acre — generally not more than one bushel. As soon as the earth is made wet enough for any chemical decomposition to take place, the oxalic acid must let go its *base* of iron, or alumina, and seize upon and combine with the lime that formed an ingredient of the gypsum. The sulphuric acid left free, will combine with the iron, or the alumina of the soil, forming copperas in the one case, and alum in the other. *The gypsum no longer exists* — and surely no more satisfactory reason can be given why no effect from it should follow. The decomposition of the gypsum has served to form two or perhaps three other substances. One of them, oxalate of lime, I suppose to be highly valuable as manure: but the very small quantity that could be formed out of one or even two bushels of gypsum, could have no more visible effect on a whole area, than that small quantity of calcareous earth, or farm-yard manure. The other substance certainly formed, copperas, is known to be a poison to soil and to plants — and alum, of which the formation would be doubtful, I believe is also hurtful. In such small quantities, however, the posion would be as little perceptible as the manure — and no apparent effect whatever could follow such an application of gypsum to an acid soil. So small a proportion of oxalic acid, or any oxalate other than of lime, would suffice to decompose and destroy the gypsum, that it would not amount to one part in twenty thousand of the soil.

Why gypsum sometimes acts as a manure on acid soils, when applied in large quantities for the space, is equally well explained by the same theory. If a handful, or even a spoonful of gypsum is put on a space of six inches square, it would so much exceed in proportion all the oxalic acid that could speedily come in contact with it, that all would not be decomposed — and the part that continued to be gypsum, would show its peculiar powers perhaps long enough to improve one crop. But as tillage scattered these little collections more equally over the whole space — or even as repeated soaking rains allowed the extension of the attractive powers — applications like these would also be destroyed, after a very short-lived and limited action.

Soils that are naturally calcareous, cannot contain oxalic acid combined with any other base than lime. Hence, gypsum applied there, continues to be gypsum — and exerts its great fertilizing power as in Loudon or Frederic.[20] But even on those most suitable soils, this manure is said not to be certain and uniform in its effects — and of course more certain results are not to be looked for with us. I have not undertaken to explain its occasional failures any more than its general success, on the lands where it is profitably used — but only why it cannot act at all, on lands of a different kind.

The same chemical action being supposed, explains why the power of profiting by gypsum should be awakened on acid soils after making them calcareous — and why that manure should seldom fail, when applied mixed with very large quantities of calcareous earth.

[H. Page 139.]

Estimates of the cost of labour applied to marling.

Before we can estimate with any truth the expense of improving land by marling, it is necessary to fix the fair cost

[20] [Virginia counties.]

of every kind of labour necessary for the purpose, and for a length of time not less than one year. We very often hear *guesses* of how much a day's labour of a man, a horse, or a wagon and team, may be worth — and all are wide of the truth, because they are made on wrong premises, or no premises whatever. The only correct method is to reduce every kind of labour to its elements — and to fix the cost of every particular necessary to furnish it. This I shall attempt: and if my estimates are erroneous in any particular, others better informed may easily correct my calculation in that respect, and make the necessary allowance on the final amount. Thus, even my mistakes in the grounds of these estimates, will not prevent true and valuable results being derived from them.

The following estimates were made in 1828, according to the prices of that year. I shall make no alteration in any of the sums, because there is no considerable difference at this time, (January 1832,) and the least alteration would make it necessary to change the after calculations founded on them. But no one estimate will suit for years of different prices. If any one wished to know the value of labour when corn (for example) was higher or lower, he must ascertain the difference in that item, and add or deduct, so as to correct the error.

Cost of the labour of a Negro Man in 1828.

Hire for the year, payable at the end, - - - - - - $38 00
Food—19½ bushels of corn at 40 cents, - - - $7 80
    Add 10 per cent. for waste in keeping, - 78
    Meat and fish, &c., - - - - - 9 00
                               ————— $17 58
Interest for one year on $17 58, paid for food, - - - 1 05
                                    ————— 18 63
Clothing—6 yards coarse three-fourths woollen cloth, at 50
    cents, - - - - - - - - $3 00
    12 yards cotton, for summer clothes and two
    shirts, at 12 cents, - - - - - - 1 44
    Blanket at $1 50, once in two years—yearly, - 75
    Shoes and mending, - - - - - - 2 50
                                  ————— 7 19

        Amount carried forward, $63 82

Amount brought forward, $ 63 8₂

Taxes—State, 47 cents—county, 47—poor, 33—road, suppose
    1 dollar, -   -   -   -   -   -   -   -   - $ 2 27
    His share of expense of quarters, fuel, and sending
      to mill, -   -   -   -   -   -   -   -   -   4 50
    Nursing when sick, (exclusive of medical aid,) -   -   1 50
                                       ———— 8 27

                                               $ 72 09
Add 20 per cent. on the whole of the above for cost of superintendence, waste, wanton damage to stock, tools, &c., and thefts,   14 41

          Total expense per year, $ 86 50
Time lost—Sundays and holydays,   58
       Bad weather and half
          holydays, -   -   -   20
       Sickness,   -   -   -   10

     From 365, deduct   88, leaves 277 working days. 277 days
costing $ 86 50, makes the cost of each working day 31¼ cents.

Remarks.

The hire was fixed at the average price obtained that year
for ten or twelve young men hired out at the highest bids,
for field labour. According to our established custom, all the
expenses of medical attendance, and loss of time from the
death of a slave occurring when he is hired, are paid, or deducted from the hire by the owner — and therefore are
omitted in this estimate. By supposing the slave to be hired
by his employer, instead of being owned, the calculation is
made more simple, and therefore more correct.

Cost of the labour of a Negro Woman.

Hire for the year,   -   -   -   -   -   -   -   - $ 10 00
Food, -   -   -   -   -   -   -   -   -   -   12 95
Clothing, blanket, and shoes,   -   -   -   -   -   -   6 50
Taxes, quarters, fuel, mill, nursing, &c.,   -   -   -   7 19
              Amount carried forward,   ————$ 37 14
              Amount brought forward, $ 37 14
Add 20 per cent. as before, for superintendence, &c.,   -   -   7 53

          Total yearly cost, $ 44 67

Suppose lost time, 108 days, leaves working days 257, at 17⅓ cents.

Nearly all the women who are usually hired out, are wanted by persons having few or no other slaves, as cooks, or for some other employment at which they are more useful, than at field labour — and their price is nearer fifteen dollars in these cases. But when there is no demand for such purposes, women for field labour will not bring more than ten dollars.

A boy of thirteen or fourteen would hire for more than the foregoing estimate of the hire of a woman, but would not lose half the time from sickness and bad weather, and therefore may be supposed to cost the same per day, or seventeen and one-third cents. A girl, fifteen or sixteen years, for similar reasons, may be put at the same price.

Cost of the labour of a Horse.

First cost of a good work horse, $ 80 00—supposed to last five years at work, makes the yearly *wear*, - - - -	$ 16	00
Interest for one year on $ 80 00—$ 4 80—Tax, 12 cents, -	4	92
	$ 20	92
20 bbls. of corn at $ 2 00—3,500 lbs. of fodder at 50 cents the hundred, - - - - - - - -	$ 57	50
Add 10 per cent. for waste in keeping, - - - - -	5	75
	63	25
Interest on $ 63 25, for one year, - - - - - -	$ 3	79
Share of yearly expense for corn-house, - - - -	47	
	4	26
Total yearly cost,	$ 88	44

Lost time, 98 days, leaves 267 working days, at 33 cents.

A mule eats less corn than a horse, but more hay, and lives longer — and may be considered as costing one-fifth less — or yearly cost — $70 00 — and daily, 26½ cents.

A tumbril for marling, will cost when new, $ 25 00		
It will last two years, or (what is the same thing) if that sum will pay for all repairs, for two years—its wear per year, is - - -	$ 12	50
Interest on $ 25 00 for a year, - - - - - - -	1	50
Cost per year,	$ 14	00

And at 267 working days—cost per day 5 cents.

In the estimate of the cost of horse labour, no charge is made for attendance, because that is part of the labour of the driver, and forms part of *his* expense. No charge is made for grazing, because enough corn and hay are allowed for every day in the year — and when grass is part of his food, more than as much in value is saved in his dry food. No charge is made for stable or litter, as the manure made is supposed to compensate those expenses.

It may be supposed that the prices fixed for corn, and fodder or hay, are too low for an average. Such is not my opinion. The price is fixed at the beginning of the year, when it is always comparatively low, because it is too soon for purchasers to keep shelled corn in bulk, and the market is glutted. Besides, the allowance for waste during the year's use (10 per cent.) makes the actual price equal to two dollars and twenty cents on July 1st. The nominal country price of corn in January, is almost always on credit — and small debts for corn are the latest and worst paid of all. The farmer who can consume any additional portion of his crop, in employing profitable labour, becomes his own best customer. The corn supposed to be used, by these estimates, is transferred on the first of January, without even the trouble of shelling or measuring, from A. B. *corn-seller*, to A. B. *marler*, and instantly paid for. Two dollars per barrel at that early time, and obtained with as little trouble from any purchaser, would be a better regular sale, than the average of prices and payments have afforded for the last eight years.

COST OF MARLING,

Founded on the foregoing estimates of the cost of labour.

From the beginning of November 1823, to the 31st of May 1824, a regular force, of two horses and suitable hands, was employed in marling on Coggin's Point, on every working day, unless prevented by bad weather, wet and soft roads, or some pressing labour of other kinds. The same two horses were used, without any change, and indeed, they had drawn the greater part of all the marl carried out on the farm, since 1818. The best of the two was seventeen years old — both of middle size, and both worse than any of my other horses, which were kept at ploughing.

The following estimates were made on a connected portion of this time and labour, and upon my own personal observation and notes of the work, from the beginning to the end. It was very desirable to me, to know the exact cost of some considerable job of marling, attended with certain known difficulties, and on any particular mode of estimating the expense: for although the same degree of difficulty, and of cost of labour, might never again be met with, still, any such estimate would furnish a tolerable rule, to apply, in a modified form, to any other undertaking of this kind. These estimates may be even more useful to other persons — as they will serve generally to prove that the greatest obstacles to the execution of this improvement, are less alarming, and more easily overcome, than any inexperienced persons would suppose.

Both these jobs were attended with uncommon difficulties, in the unusual thickness of the superincumbent earth, compared to that of the fossil shells worth digging, and on account of the distance, and amount of ascent to the field. The first job was so much more expensive than was anticipated, that it may perhaps be considered as a failure — but as the account of its expense had been kept so carefully, it will be given, just as if more success and profit had been obtained. This work was commenced April 14th, 1824. The bed of marl for the upper six feet of its thickness, is dry and firm, though easy to dig, and rich. It has an average strength of $\frac{45}{100}$ — the shells mostly pulverized, and the remaining earth more of clay than sand. After being carried out, the heaps appear, to a superficial observer, to be a coarse loose sand. Below six feet, the marl became so poor as not to be worth carrying out, and was not used except when the distance was very short. Its strength was less than $^2\%_{00}$. The bed at first was exposed on the surface, near the bottom of a steep hill-side — but as a large quantity had been taken out, and several successive cuts made into the face of the hill, some years before, the covering earth was increased, on the space now to be cleared, so as to vary between eight and sixteen feet, and I think averaged between eleven and twelve. The situation of the marl and road required that a clear cartway should be made as low as the intended digging: and therefore nearly all of the earth was to be moved by a scraper, and was thrown into the narrow bottom at the foot of the hill. This earth, served thus to form an excellent causeway across the valley, which made part of the road in the next undertaking. All this marl runs horizontally, and the layers of different qualities are very uniform in their thickness. The greater part of the covering earth is a hard clay, or impure *fuller's earth*, which was difficult to dig, and still more so for the scraper to take up and remove. Part was thrown off by shovels, and served to increase a mound made by former operations, within the circle around which the scraper was carried.

Labour used in digging and removing earth.

4 days' labour of 9 men,	at 31¼ cents each, -	-	-	-	-	-$ 11	25
4 6 women, } at 17⅓ cents. -	-	-	-	-	-	5	58
4 2 boys, }							
4 1 young girl at 15, and 1 old man at 25,	-	-	-	1	60		
3 8 oxen, (the scraper being drawn by 4 half the day, which then rested and grazed while the others worked the other half of the day,)—at 20 cents each, -	-	-	-	-	-	4	80
Add 80 cents for wear of scraper, hoes, and shovels,	-	-	-	-		80	
Total,						$ 24	03

The price allowed for the oxen is much too high for common work, and so much rest allowed: but they work so seldom at the scraper, that both the men and the oxen are awkward, and the labour is very heavy, and even injurious to the team.

Labour of digging and carrying out the Marl.

Three tumbrils were kept at work on this job and the next, a good mule being added to the regular carting force — and no time was lost from April 20th, to May 31st, except when carts broke down, (which was very often, owing to careless driving, and worse carpentry,) or when bad weather compelled this labour to stop. One man dug the marl and assisted to load: another man loaded, and led the cart out of the pit, until he met another driver returning from the field, to whom he delivered the loaded cart and returned to the pit with the empty one. Of the two other drivers, one was a boy of sixteen, and the other twelve years old — the youngest only was permitted to ride back, when returning empty. The distance to the nearest part of the work (measured by the *chain*,) was nine hundred and two yards, and the farthest one thousand and forty-five: adding two-thirds of the difference to the nearest for the average distance, makes nine hundred and ninety-seven yards. The ascent from the pit, by a road formerly cut and well graduated, for marling, was supposed to be twenty-five feet in perpendicular height — and every trip of the carts, going and coming, crossed a valley, supposed to be fifteen feet deep, and both sides forming a hill-side of that elevation.

When only four and a half feet of the marl had been dug, a large mass of earth fell into the pit, covered entirely the remaining one and a half feet of marl, and stopped all passage for carts. To clear away this obstruction would have cost more labour than the remaining marl was worth, and therefore this pit was abandoned. This happened on May 10th, when six hundred and ninety-nine loads had been carried out, and the work done was equal to thirty-six days' work of one cart (by adding together all the working time of each) — which was nineteen and a half loads for the average daily work of each cart, or fifty-eight for the three. The average size of the loads, by trial, was five and a half heaped bushels — and the weight, one hundred and one pounds the bushel. It was laid on at one hundred and four loads or five hundred and seventy-two bushels the acre.

Labour employed, for 669 loads, or 3680 bushels.

2 men at 31¼ cents, - - - - - - -	62½
2 boys at 19 cents, - - - - - - -	38
2 horses at 33 cents, - - - - - - -	66
1 mule at 26½ cents, - - - - - - -	26½
3 carts at 5 cents—tools at 3, - - - - -	18

Daily expense, or for 58 loads,	$2 11

Digging and carting 699 loads at the same rate, - - - -	$25 25
Add the total expense of removing earth, - - - - -	24 03

	$49 28
Spreading at 31¼ cents the 100 loads, - - - - - -	2 19

Total expense,	$51 47

Which makes the cost per bushel,	1 34–100 cents.
per load, (5½)	7 36–100
per acre, of 572 bushels, $7 66	

This marl was laid on much too thick for common poor land, and one-fourth of the body uncovered was lost, by the falling in of the earth. If one-fourth of the expense of uncovering the marl, was deducted on account of this loss, it would reduce the whole expense nearly one-eighth.

As soon as the carts were stopped in the work just described, they were employed in moving earth from similar marl, across the ravine. The thickness, strength, and other qualities of the marl, on both sides, are not perceptibly different. A large quantity had also been formerly dug on this side, but the land being lower, the covering earth was not more than ten feet where thickest, and the average was eight and a half or nine feet. To make room for convenient working, and a large job, an unusual space was cleared, ten to fourteen feet wide, and perhaps fifty or more long. The shape of the adjoining old pits, compelled this to be irregular. The greater part of the earth was of the same hard fuller's earth mentioned as being on the other side — and the upper part of this was still worse, being in woods, and the digging was obstructed by the roots and trees.

Labour used in digging and removing the earth.

6 men	4 days,	} at 31¼ cents, - - - - -	$ 8	43¾
1 man	3	}		
5 women	5	}		
1 woman	1	} at 17⅓ cents, - - - - -	6	24
2 boys	5	}		
1 old man	2	25 cents, - - - - -		50
2 girls	6	15 cents, - - - - -	1	80
8 oxen, for the scraper, as before, each team at rest half the day,				
	5 days,	at 20 cents, - - - - -	8	00
3 horses and carts, 1½		at 38 cents, - - - - -	1	71
Add for damage to scraper and other utensils, - - - -				80

Total cost of moving earth,	$ 27	48¾

Enough of the earth was carried by the carts to the dam crossing the ravine, to raise the road as high as the bottom of the intended pit. The balance was thrown into the valley whenever most convenient. Only a small proportion, perhaps one-third, could be thrown off, without being carried away by the carts and scraper.

The loads were carried to the same field, and by the same road as from the former digging. The first hundred and ninety-one loads served to finish the piece begun before, of which the average distance was nine hundred and ninety-seven yards: all the balance was carried to land adjoining the former, eight hundred and forty-seven measured yards from the pit.

The loads were ordered to be increased to six bushels, which was as much as the carts (without tail-boards) could hold, without waste, in ascending the hills: but as the loaders often fell below that quantity, I suppose the average to have been five and three-fourths heaped bushels, or five hundred and eighty-one pounds.

The tumbrils were kept constantly at this work, except when some of the land was too wet, or for some other unavoidable cause of delay. All the space which the old pits occupied, and over which the road passed, being composed of tough clay thrown from later openings, and which had never become solid, was made miry by every heavy rain, and caused more loss of time than would usually occur at that season. The same four labourers, and two horses, and one mule, employed as before — and their daily work was as follows. —

May	13th, Began the new pit—2 tumbrils all the day, and 1 for 2 hours only, afterwards otherwise employed, - - - - - - -	47 loads.
	14th, 2 tumbrils half the day, then employed otherwise—(1 horse idle,) - - - - -	21
	15th, 3 ditto, - - - - - - -	61
Monday 17th, 3 ditto, finished most distant work with -	62
		191
	and began nearest with - - - -	4
	18th, 3 ditto, for 4 hours (stopped by heavy rain,)	22
	19th and 20th, 3 ditto, carrying manure, on drier land,	
	21st, 3 ditto, return to marling, - - - -	75
	22d, Rain—no work done by horses.	
Monday 24th, 1 ditto, moving manure.	
	25th, 3 ditto, return to marling, - - - -	74
	26th, 3 ditto, ditto, - - - - -	75
	27th, 3 ditto, ditto, - - - - -	72
	28th, 3 ditto, ditto, - - - - -	72
	29th, 3 ditto, (shafts of one broken and repaired,) - - - - - - -	64
Monday 31st, 3 ditto, until rain at 4, P. M. - - -	53
		511
		702

average distance 997 yards.

average distance 847 yards.

After this stoppage, the horses were put to ploughing the corn, that the cultivation might be sufficiently advanced to use all the labourers in harvest, which began on the 11th of June. As near as I could determine by inspection, and a rough cubic measurement, about one half of the uncovered marl was then dug and carried out. As the remainder was not dug until August, when I was absent from home, I have no more correct means of ascertaining these proportions; and shall according to this supposition charge *half* the actual cost of the whole uncovering of earth, to this supposed half of the marl which formed this last operation.

The list of day's work shows that the average number of loads per day, at eight hundred and forty-seven yards, was twenty-four and a half for each cart, which made twenty-three and a half miles for the day's journey of each horse. The first four days' work finished the farthest piece, of which the average distance was nine hundred and ninety-seven yards — but this part of the work was on the nearest side of that piece, and at less than that average distance. I shall not make any separate calculation for these hundred and ninety-one loads, but consider all as if carried only eight hundred and forty-seven yards.

The daily cost of the labouring force, 2 men, 2 boys, 2 horses, and 1 mule, was before estimated at $2 11—which served to carry out 73½ loads, or 422 bushels. At that rate, (to May 31st,) 702 loads, or 4036 bushels, cost,   -   -   -   -   -   -   - $ 20   15
Add half the expense of uncovering, (half the marl still remaining not dug,)   -   -   -   -   -   -   -   -   13   74
For spreading, at 31¼ cents per hundred loads,   -   -   -   2   18¾

Total cost of 4036 bushels laid on, $ 36   07¾

Which makes the cost per bushel, 9 mills nearly.
And per acre, at 104 loads, or 598 bushels,   -   -   -   - $ 5   34½
Or at 400 bushels, which would have been a sufficient, and much safer dressing,   -   -   -   -   -   -   -   - $ 3   57½

In 1828, at Shellbanks, a very poor, worn, and hilly farm, I commenced marling, and in about four months, finished one hundred and twenty and a half acres at rates between two hundred and thirty and two hundred and eighty bushels per acre. The time taken up in this work, was five days in January, and all February and March, with two carts at work

— and from the 5th of August, to the 27th of September, with a much stronger force. I kept a very minute journal of all these operations, showing the amount of labour employed, and of loads carried out during the whole time. It would be entirely unnecessary to state here any thing more than the general amounts of labour and its expense, after the two particular statements just submitted. At Shellbanks, the difficulties of opening pits were generally less — the average distance shorter, and the reduced state of the soil, and the strength of the marl, made heavy dressings dangerous. These circumstances all served to diminish the expense to the acre. The difficulties, however, at some of the pits, were very great, owing to the quantity of water continually running in through the loose fragments of the shells — and almost every load was carried up some high hill. Taking every thing into consideration, I should suppose that the labour and cost of this large job of marling will be equal to, if not greater, than the average of all that may be undertaken, and judiciously executed, on farms having plenty of this means for improvement, at convenient distances.

Cost of marling 120½ acres at Shellbanks.

Preparatory work, including uncovering marl, cutting and repairing the necessary roads, and bringing corn for the team—Digging, carrying out, and spreading 6892 loads (4½ heaped bushels) of marl, $ 250   38
At the average rate of 57½ loads, or 259 bushels per acre, the average expense was, to the acre,   -   -   -   -   -   -   -   -   2   08
To the load,   -   -   -   -   -   -   3 cents and 63-100ths,
And to the bushel,   -   -   -   -   -   0   83-100ths.

FINIS.